T0276899

STRONG ANTICIPATION

Compensating Delay and Distance

STRONG ANTICIPATION

Compensating Delay and Distance

Susie Vrobel

The Institute for Fractal Research, Germany

World Scientific

NEW JERSEY • LONDON • SINGAPORE • BEIJING • SHANGHAI • HONG KONG • TAIPEI • CHENNAI • TOKYO

Published by

World Scientific Publishing Co. Pte. Ltd.
5 Toh Tuck Link, Singapore 596224
USA office: 27 Warren Street, Suite 401-402, Hackensack, NJ 07601
UK office: 57 Shelton Street, Covent Garden, London WC2H 9HE

British Library Cataloguing-in-Publication Data
A catalogue record for this book is available from the British Library.

STRONG ANTICIPATION
Compensating Delay and Distance

ISBN 978-981-12-8198-3 (hardcover)
ISBN 978-981-12-8199-0 (ebook for institutions)
ISBN 978-981-12-8200-3 (ebook for individuals)

For any available supplementary material, please visit
https://www.worldscientific.com/worldscibooks/10.1142/13560#t=suppl

Typeset by Stallion Press
Email: enquiries@stallionpress.com

Printed in Singapore

Preface

My favourite topics — the nature of time and the Now — have preoccupied my life since the early 1980s and resulted in my Theory of Fractal Time. A general interest in fractals and chaos theory routed me towards Otto Rössler's micro-relativity and endo-perspectives. His notion of the Now as an interface brought me full cycle to where I got started: the fractal Now. From there on, it was only a small step to the concept of anticipation I encountered in Daniel Dubois' CASYS conferences. Since then, I have talked and written about the many aspects of anticipation, mostly in connection with fractal time and observer extensions. My approach has always been systems-theoretic and phenomenological. This book is no exception and comprises my research of the last thirty years in this area. I hope you will enjoy this book and would be most grateful for feedback.

Susie Vrobel
Bad Nauheim
July 2023

Acknowledgements

Many thanks to my husband Barry Baddock for supporting all my research projects and enduring my follies, for providing helpful hints while I was writing this book and brushing up on the manuscript. I am also indebted to Otto Rössler for lasting encouragement and support, to Daniel Dubois for launching and steering his CASYS conferences and to George Lasker for organizing IIAS. Over the years, these conferences have been most pleasant and inspiring events. I am grateful to the many friends and colleagues for past and present inspiring discussions during these meetings. I would also like to express my thanks to the participants of my stream 'Models of Embodied Cognition' at the OR Conference in Vilnius 2002. Many thanks, too, to Gerd Altmann for providing the cover picture.

Contents

Chapter 1

Introduction: Predicting the Future

The idea that future events influence the present is not new. But rather than waiting for the future to unfold, mankind has tried to catch a glimpse of things to come with the help of omens and other harbingers of good and bad news. Predicting the future took (and still takes) unusual forms such as the Assyrian divination from sheep guts, Turkish coffee readings and spider divination in Cameroon. In Europe, comets, eclipses and unnatural births used to be harbingers of bad news (see Fig. 1.1). Today, black cats and broken mirrors have taken their place, but a predisposition to omens remains, providing fertile ground for abuse and self-fulfilling prophecies.

For the above examples, there is no causal chain between a present omen and the future event it predicts (if divine intervention is not regarded as an option). However, now and then, strong correlations make a convincing case, waiting to be proven or disproven [1]. And the claim that no causal connection whatsoever exists is impossible to prove, as there may always be hidden connections.

Observed correlations often appear inexplicable at first sight until a causal chain is revealed. An example of such a correlation can be found in the case of the UNHCR's data scientist whose task was to predict the number of displaced people from Somalia arriving at a refugee camp in Ethiopia. He found out that the price of goats in Somalia predicted the number of refugees the camp in Ethiopia could expect. The explanation for this correlation is not obvious and took the scientist a while to

Fig. 1.1. Examples of omens from the Nuremberg Chronicle (1493): Natural phenomena and unnatural births [3].

figure out: Before people left, they would sell their goats, as they wouldn't survive the journey, which led to excess supply of goats in Somalia and a fall in price [2].

The revealed causal connection would allow the data scientist to establish a model to predict the influx of migrants. This modelling relation could be written down, formalized and used by everyone: An external model to predict future migration.

Examples of external models are those which anticipate the influx of migrants or climate models which predict the formation of tropical cyclones.

Other predictions are based on an internal, wired-in model, such as neural assemblies in our brain which are pre-activated by our auditory system when we anticipate a musical note. Our present behaviour may also be determined by future states of an environment with which we are coupled, directly or indirectly, such as the daily light–dark cycle which entrains our circadian rhythms. And finally, there is the coordination between internal and external long-term correlations with their nonlocal relations and prediction through advanced signals. Anticipation relies on models or exists as a model-free architecture; it can be weak or strong, local or global.

This book is concerned with both local and global strong anticipation, which manifests itself either as a form of complexity reduction through delay and distance compensation or through the coordination of multi-layered long-term correlations.

In a phenomenological and systems-theoretical approach, I shall show how descriptions of anticipative behaviour are based on underlying assumptions about the nature of space and time as well as our situation in and perspective of the world within and around us. Some of these assumptions are in plain view; others are hidden, or transparent. This book sets out to reveal such transparency as the result of compensatory acts and to provide a model of time to describe strong anticipation.

Chapter 2 looks at prominent constraints which limit our description of the world within and around us, such as the modelling relation. Model-based approaches, from Rosen's rejection of the reactive paradigm to predictive processing, are discussed. Nonmodel-based architectures such as the embodied and enactive paradigm are dealt with against the background of the question as to whether there is a need for representation.

The notion of anticipation is introduced and differentiated between weak and strong anticipation, following Dubois' definitions, on which Stepp and Turvey base their definition of anticipatory synchronization. Kelty-Stephen and Dixon also adopt Dubois' notions of weak and strong anticipation and make a further distinction between local and global strong anticipation: Local strong anticipation (which they equate with anticipatory synchronization) results from the coupling of organism and environment. Global strong anticipation, on the other hand, unfolds simultaneously on multiple time scales and appears to result from the coordination of long-term correlations of internal and external fractal structures. The role of fractal scaling in strong anticipation was also emphasized by Dubois and is implicit in his notion of hyperincursion.

Finally, our endo-perspective is presented as a constraint on the obserpant's (= observer-participant) perceptions, theories and models of the world. Our primary experience of time and space is an example of underlying assumptions concealed in our theories and models, which are secondary constructs and thus anthropocentric.

Chapter 3 introduces the notion of boundary shifts between agent and environment. The question as to when a systemic whole displays strong anticipation is discussed against the background of spatial and temporal extensions of obserpants and robotic devices. Spatial and temporal obserpant extensions and reductions can be compensated through the recalibration of obserpant–world interfaces.

The four conditions obserpants need to meet in order to qualify as local strong anticipative systems are as follows: the occurrence of a boundary shift towards the outside, the boundary shift results in the formation of a new systemic whole, the interfaces which have been merged or have newly emerged are transparent to the obserpant, and the obserpant retains a sense of agency after the boundary shift and spatio-temporal compensation. In contrast to Stepp and Turvey, as well as Stephen, Dixon and Stepp, who assign representations and models strictly to weak anticipation, my four conditions for strong anticipation do not exclude representations. They also introduce a new condition, i.e. that the obserpant needs to retain a sense of agency.

Chapter 4 focuses on the various manifestations of strong anticipation as delay and distance compensation. The phenomenon known as anticipatory synchronization emerges from enslaved systems synchronizing with a master system. Other delays are less visible yet act as natural constraints to our actions and cognition. Neurocognitive delays are add-ons, which need to be taken into account before any interaction with the rest of the world. Compensating neurocognitive delays may manifest themselves not only as anticipation but, alternatively, also as postdiction (or both, which Eagleman denotes as peri-diction).

Transparency is another such add-on to be considered. Metzinger's Phenomenal Self Model reveals transparency, i.e. the fact that we are not aware of the representational character of our experience, as a selection effect which allows us to skip unnecessary delays and makes our interaction with the world more immediate. This skipping is a compensatory act and thus another example of strong anticipation.

The concept of a simultaneity horizon as defined by Pöppel is another constraint which defines a systemic whole as a system whose components share a boundary within which multi-modal impressions are perceived as occurring simultaneously.

Examples of strong anticipation in physical and biological systems follow. The status of predictive homeostasis and epigenetics is discussed against the background of strong anticipation. The chapter closes with a new approach by Stepp and Turvey, Delignières and Marmelat, and Washburn, who present anticipatory synchronization in human interaction as bi-directional coupling rather than a master–slave relationship.

Chapter 5 addresses the need for a model of time which can describe anticipatory systems. Rosen, Nadin, Poli and others have dealt with this issue, differentiating between the notions of succession and simultaneity. However, in order to describe not only succession but also nested simultaneity in multi-layered phenomena, which are the basis for local and global strong anticipation, an appropriate model needs to include the notion of temporal and spatial fractal structures.

My Theory of Fractal Time provides, with its temporal dimensions Δt_{depth}, Δt_{length} and $\Delta t_{density}$ (measuring nested simultaneity, succession and the fractal dimension of time), a model which is suitable for describing anticipative systems. Delay compensation in the form of strong anticipation becomes formally describable in terms of Δt_{depth} and Δt_{length}: Whenever a delay is compensated, Δt_{length} is transformed into Δt_{depth}.

Notions of strong anticipation, be they local or global, model-based or model-free, based on coupling or on the coordination of long-term correlations, are all founded on presuppositions about the nature and structure of time. However, these are secondary products based on our primary perception of time, which we experience as duration, simultaneity and succession. My Theory of Fractal Time takes account of these primary perceptions of time and provides a suitable framework to describe anticipatory systems.

Chapter 6 gives a brief introduction to notions of embodied cognition and embodied anticipation. The prevailing definition of embodiment comprises three conditions: The agent is to be situated, extended and distributed. The assignment of representations to weak anticipation and architectures without representations to strong anticipation (and direct perception) are relativized. The concept of strong embodiment allows for representations, as long as the body is given a clear explanatory role. Analogously, embodied anticipation is defined as compensatory action which is explained by bodily causes but allows for representations, including those of the body. This makes embodied anticipation an example of strong anticipation as witnessed, for instance, in the transformation of extrapersonal into peri-personal space.

The last part of Chapter 6 revisits the topic of anticipation in artificial agents and assesses how far embedded and situated robots can be denoted as strong anticipative systems.

Chapter 7 focuses on Winnicott's notions of transitional objects and potential spaces, which function as precursors to strong anticipation, as they provide the intermediate space between obserpant and environment necessary for an obserpant extension. Transitional objects live in potential spaces and come not only in the form of teddy bears or cuddly blankets but also, later in life, in the shape of less tangible phenomena like art or religion.

The intermediate space with its transitional objects belongs neither to the obserpant nor to the environment — it partakes in both and thus connects and separates self and nonself. It therefore performs the function of an extended interface.

Once transitional objects have become obsolete, they are compensated and discarded. Their meaning and function, however, survive as internalized structures within the obserpant. The compensation of transitional objects is a form of strong anticipation: In the act of compensation, the obserpant and the transitional object form one systemic whole which comprises the potential space. Once the structures and meaning of the object which previously resided in potential space have been internalized, a boundary shift towards the outside has occurred. This interfacial shift is transparent to the observant, yet he or she retains a sense of agency after recalibration.

Chapter 8 introduces the notion of embodied trust as an obserpant extension. Some such extensions imply a modification in our sense of proprioception and increase peripersonal space. Physical extensions, like a healer's touching hand, change our muscle tone, which induces a feeling of trust. But trust also emerges from logical reasoning, from adopting abstract frameworks which are mediated via symbol systems.

Two types of trust are differentiated. The first, symbolic trust, results from a reduction of external complexity through symbol systems. It is based on digital coding and manifests itself as weak anticipation. The second type, embodied trust, results from a reduction of external complexity through physically compensated delays or distances. It is based on analogue coding and manifests itself as strong anticipation. However, if an interpreting obserpant contextualizes and thus embodies digital coding, he or she transforms weak into strong anticipation, which eventually evolves into habits.

Chapter 9 introduces the notions of fractal and nonfractal obserpants and looks at the benefits and dangers of contextualization and decontextualization. The relationship between internal and external temporal structures is defined as Interface Complexity (IC). For local strong anticipation, IC is measured as the relation between embodied and embedding parameters: IC = Δt_{depth} (obserpant)/Δt_{depth} (environment). For global strong anticipation, IC is measured in the fractal dimension as the relation between embodied and embedding long-term correlations: IC = $\Delta t_{\text{density}}$ (obserpant)/ $\Delta t_{\text{density}}$ (environment), where the environment can also consist of another obserpant.

Chapter 10 presents local strong anticipation as a blind spot, as it conceals hidden compensatory acts. As such, it presents a constraint to any epistemological endeavour. But we have tools to reveal compensated temporal and spatial extensions we did not even suspect existed. They can be revealed if the endo- and exo-perspectives differ in terms of the temporal dimensions perceived. What is compensated to the endo-obserpant may be uncompensated from the exo-perspective. The difference reveals compensatory acts as phenomenal blind spots. A new kind of relativity contrasts compensated and uncompensated delays and distances as perceived from endo- and exo-perspectives. Against this background, local strong anticipation acts as a blind spot that may evolve, first into regularities and habits and eventually into natural laws.

Chapter 11 deals with dynamical diseases as manifestations of compensated and uncompensated delays. Local strong anticipation results from compensating spatio-temporal extensions and synchronizing with two or more environmental systems. Global strong anticipation, by contrast, results from the coordination of long-term correlations of fluctuations on multiple time scales on both sides of the interfacial cut.

While local strong anticipation can lead to both healthy and pathological states, global strong anticipation is generally regarded as being conducive to good health. Pink noise ($1/f$ noise), which exhibits fractal long-term correlations, is a sign of healthy dynamics and an ideal degree of complexity. Deviations from pink noise tend to cause pathological states. Morbus Parkinson, a neurodegenerative disorder, is a typical example of multiple deviation towards Brownian noise which shows itself in an increasing rigidity and a loss of complexity in gait and speech. A deviation towards white noise has been observed in the heartbeat in atrial fibrillation and gait in Huntington's disease.

If the formation of Δt_{depth} is compromised, both positive and negative effects can result. A decline in Δt_{depth} is generally detrimental to health, as is apparent in Morbus Parkinson and Depression. However, it can be helpful in situations where focusing is of vital importance. By contrast, a lack of or reduction in $\Delta t_{\text{density}}$ tends to cause dynamical diseases or disorders.

A transformation of Δt_{length} into Δt_{depth} increases interfacial complexity and is conducive to health. By contrast, turning Δt_{depth} into Δt_{length} results in a loss of complexity, which reduces the range of possible responses and signifies a reduction in global strong anticipation. Thus, a shift in the temporal dimension from Δt_{depth} to Δt_{length} can be an indicator of dynamical diseases or disorders.

Chapter 12 adds the phenomena of nonlocality and insight to examples of strong anticipation. Compensating delay and distance in spacetime generates local strong anticipation through the coupling of organism and environment (or two organisms). Wheeler's delayed choice may point to the removal of an existing delay.

An example of macroscopic nonlocality is Korotaev et al.'s results of their Baikal experiments. The resulting nonlocal correlations are of a probabilistic and global nature and show nonlocal connections, which Einstein referred to as "spooky action at a distance". They are interpreted as global strong anticipation.

A philosophical path to strong anticipation lies in the acquisition of insight, as outlined by philosopher Jiddu Krishnamurti. According to him, insight is achieved when all inward movement ceases. Then the internal and external worlds are no longer separated and we can partake in a universal mind and an eternal Now.

However, the cessation of internal movement is an impossible state to achieve. The internal noise of the obserpant cannot be eradicated and may even be necessary in order to create a state in which all succession ends. This notion of insight corresponds to a fractal condensation scenario which generates sheer simultaneity, with Δt_{depth} approaching infinity and Δt_{length} ceasing into (almost) nothingness, as there would be no more succession.

Nottale's and El Naschie's notions of fractal spacetime belong both to the realms of phenomenology and ontology. On the one hand, the fractal geodesics of spacetime constrain, as an ontological entity, the obserpant's and environmental microscopic movements. On the other hand, embodied brains have developed the concept of fractal spacetime as a result of our co-evolution with a phase of the universe. Phenomenology and ontology are inextricably interwoven in our interfaces because, as obserpants, we simultaneously partake in both the quantum and classical levels with which we co-evolved.

At the bottom line, strong anticipation, be it in the form of delay and distance compensation, the coordination of long-term correlations, insight or nonlocal perspectives, is a Janus-faced sword. On the one hand, it allows us to steer with confidence our bodies (including our brains) and extensions thereof through our environment. On the other hand, it makes us blind to existing compensated delays and distances or to long-term correlations and thus acts as an epistemological constraint. Revealing our unconscious compensatory acts may eradicate our phenomenal and epistemological blind spots but comes at a price — the end of naïve realism and smooth navigation.

References

[1] J. L. Casti, *Mood Matters — From Rising Skirt Lengths to the Collapse of World Powers*, Springer, 2010, p. 202.

[2] A Goat Story, *UNHCR Innovation Service*, May 8, 2019. https://medium.com/unhcr-innovation-service/a-goat-story-3ed6bdd2b237.

[3] Illustrations from *The Nuremberg Chronicle*, by Hartmann Schedel (1440–1514), original in colour, public domain. https://de.wikisource.org/wiki/Schedel%E2%80%99sche_Weltchronik.

Chapter 2

Models, Weak and Strong Anticipation

When we intend to talk about an abstract notion such as anticipation, we presuppose, consciously or unconsciously, more basic concepts which are more or less precisely defined in our models, theories and paradigms. The modelling relation between natural and formal systems rests on logical shortcomings which need to be addressed. Encoding properties of a natural system into a formal system and decoding them by projecting formal relations onto the natural system entail a logical mix-up: Logical implications in the formal system are equated with causal relations in the natural one. This way, predictions are made which muddle up formal and material causes. Adding a final cause to the equation takes account of the natural system's inherent dynamics towards a purpose, a future state.

In addition to the modelling relation, there are other constraints to be considered when we describe the dynamics of a complex system, such as anticipative behaviour. Mathematician and philosopher Norbert Wiener was well aware that we have to make compromises when we model the world in and around us. He remarked that "the best model of a cat is a cat, preferably the same cat" [1] but conceded that this ideal is not achieved in practice. We must therefore contend with abstractions and focus on selected parameters which we may then consider within a reasonable period of time. This compromise feeds into our inherent desire for complexity reduction and is sufficiently precise for most purposes in our everyday lives.

But there are other compromises we have to face when we attempt to describe the world in and around us with the help of a model: We have to take account of the fact that our models are based on preconceptions about space, time and causality, that our access to the world is mediated through interfacial cuts, and our models are created in active involvement with our environment. In short, our models are anthropocentric. Therefore, also models which describe anticipative behaviour are based on preconceived paradigms of the scientific community and individual intentions.

2.1 Rejecting the Reactive Paradigm

A model in which present behaviour is determined by an anticipated future was described by theoretical biologist Robert Rosen, the pioneer of anticipative systems. He mused that he would gladly "vacate the premises" if he saw a bear appearing on his path. This reaction would be based on a model which hints at the dire consequences of such an encounter. He concludes:

> I thus change my present course of action, in accordance with my model's prediction. Or, to put it another way, my present behavior is not simply *reactive*, but rather is *anticipatory*. [2]

Rosen suggested that anticipatory behaviour is not limited to conscious beings capable of learning. He describes the behaviour of negatively phototropic primitive organisms, which are drawn towards darkness because darkness is correlated with a moist environment and fewer predators. Rosen concludes that a wired-in model correlates darkness with an advantageous environment for those organisms, which then change their present behaviour on the basis of that model's prediction of the future [3].

He rejected the reactive paradigm, i.e. the idea that a system merely responds to a stimulus, in favour of an anticipatory conception based on the notion of a predictive internal model:

> ... an anticipatory behavior is one in which a change of state in the present occurs as a function of some predicted future state, and (...) the

> agency through which the prediction is made must be, in the broadest sense, a model. [4]

Rosen realized the shortcomings of the modelling relation between natural and formal systems. For instance, the fact that modelling relations are based on a number of presuppositions such as synonymy:

> We are going to force the name of a percept to be also the name of a formal entity; we are going to force the name of a linkage between percepts to also be the name of a relation between mathematical entities; and most particularly, we are going to force the various temporal relations characteristic of causality in the natural world to be synonymous with the inferential structure which allows us to draw conclusions from premises in the mathematical world. (...) In short, we want our relations between formal and natural systems to be like the one Goethe postulated as between the genius and Nature: what the one promises, the other surely redeems. [5]

He saw the modelling relation between formal and natural systems as logically inconsistent because inference in formal systems is translated into causality in natural systems (see Fig. 2.1). Furthermore, he emphasized the need to include contextual causes:

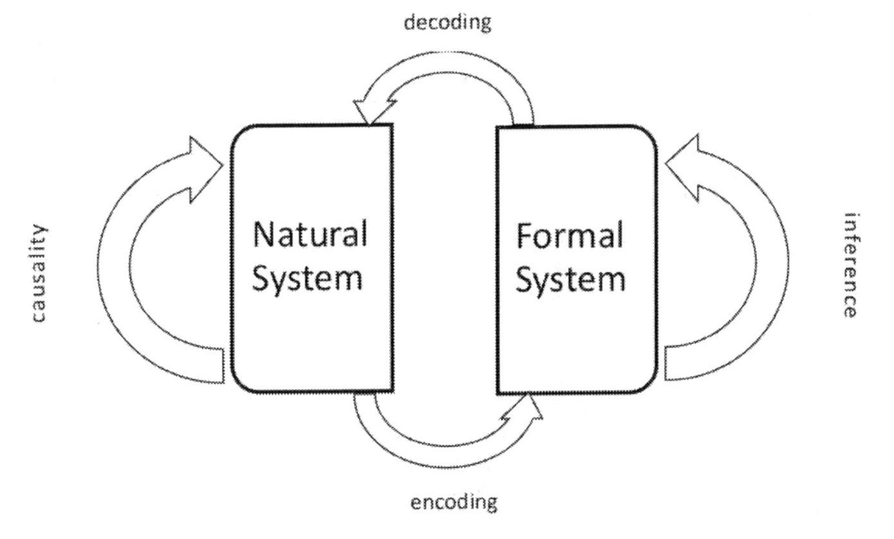

Fig. 2.1. The modelling relation.

> ... systems can establish the context for other systems and therefore formal cause can vary, ensuring that no single descriptive formalism can be complete. [6]

Rosen was also acutely aware of the fact that the separation between observer and observed leads to epistemological tangles and criticized the mechanistic view:

> ... because it uses a natural system — ourselves — to describe nature as a set of external events, which then appear to belong to a universe that has been separated from the context in which we perceive it. [7]

Systems are contexts for other systems, and although the nesting of contexts may expand *ad infinitum*, the best way to describe an anticipative system is in context, i.e. in interaction with its environment.

2.2 Action in Perception: Internal and External Constraints

The spatio-temporal extension of our bodies, including physical extensions thereof, participate in our perception and cognition of the world. Philosopher and cognitive scientist Alva Noë stresses that consciousness is not limited to the brain but includes the dynamics and makeup of our body and our environment. We enact our perceptual experience:

> Perception is not something that happens to us, or in us. It is something we do. [8]

Like a blind person perceives his or her immediate environment, we perceive the space around us, step by step, moving and touching and colliding with objects:

> Perceiving how things are is a mode of exploring how things appear. How they appear is, however, an aspect of how they are. To explore appearance is thus to explore the environment, the world. To discover how things are, from how they appear, is to discover an order or pattern

> in their appearances. The process of perceiving, of finding out how things are, is a process of meeting the world; it is an activity of skillful exploration. [9]

In this enactive approach to perception and consciousness, constraints limit both the agent and environment: The agent's physical abilities, such as the ability to move in space and agility of the limbs, encounter environmental constraints:

> Consider that to feel a table — to learn about it by touch — is to encounter it in such a way that one's movements are, in appropriate ways, impeded by the table. In general, one might say, to feel a shape or texture of a surface is, in this way, to allow one's movement to be molded by that which one touches. (...) The roundness, of course, exists apart from how it affects the probing hand. But for something to feel round (...) is precisely for it to affect the movement of the probing hand in a family of related ways. [10]

Noë stresses the importance of regarding a system and its environment as one systemic whole, which means not only does the whole display properties which cannot be found in the elements which constitute it, but it also implies that it makes no sense to regard a system in isolation when we try to describe perception and consciousness:

> Given that, as a matter of fact, we spend our lives in tight coupling with the environment (and other people), one can reasonably wonder why we find it so plausible that there could be a consciousness like ours independent of active exchange with the world. [11]

A blind person using a white cane to perceive his or her environment will behave like one systemic whole made up of him- or herself and the white cane. The blind person will perceive the world through the cane and, after a time of adaptation, that stick will feel almost like an extension of a limb. The extension of the stick itself is compensated if the spatio-temporal gap it covers has become transparent to that person, i.e. he or she is no longer aware of it.

Action-in-perception includes this type of compensation which covers our physical extensions, including our sphere of influence, when we interact with our environment. By bridging, i.e. compensating, the spatio-temporal extension which spanned between the "unextended" person and the extended one, we skip a spatio-temporal extension which used to be part of our environment but is now incorporated. As we see in Chapter 3, compensating spatio-temporal extension manifests itself as either anticipation or postdiction.

2.3 Embodied and Enactive Approaches: Questioning the Need for Representation

Rosen's notion of anticipation is representational and based on an internal model. By contrast, Noë's action-in-perception approach is nonrepresentational. Opinions on whether there is a need for representative models are divided. In fact, there is a growing community which discards models altogether.

If perception and cognition could be explained without representation, one compensatory step would be eliminated. Proponents of nonrepresentative approaches include the Embodied and Enactive Cognitive Sciences (EECS), who reject the idea of neural representation and the claim that cognition can be reduced to mental representations [12]. Early enactivists reject the idea that representations exist in the brain altogether. However, they acknowledge their existence in models as representations of the scientist's activities.

Psychologist Michael Turvey addresses the difficulty of assuming the existence of representations by pointing out the intricacy of trying to reconcile direct and indirect perception. The former does not entail any form of representation, whereas the latter does. An example is a creature walking over a surface, which is a relation between the walking organism and the surface which supports it:

> ... the supporting surface is as essential to posture and locomotion as are, for instance, the organism's legs. (...) in order to control posture and locomotion, the organism must direct its perceiving towards that same

> surface (its perceiving must be about walking on the surface). Thus, it would seem that a two-term relation involving the same surface or ground can exist in both cases: the organism walks on the ground and the organism perceives the ground … [13]

If, for the walking creature, another object in the form of a representation exists between it and the surface, perception is a three-term relation, i.e. indirect perception:

> The content of the organism's perception, what the perception is about, is the surrogate of the surface, not the surface. [14]

The difficulty which arises in the three-term relation is that very crudely, organisms do not stand or run on images or ideas or representations [15].

However, if there is no other object (no representation) between the walking creature and the surface, perception is a two-term relation, i.e. direct perception. Turvey suggests that direct perception and strong anticipation may be identical concepts, at least that strong anticipation can be seen as a generalization of direct perception:

> … each rejects a predicting model, each promotes lawful relations, and each mandates that the organism and the environment be taken together as a single system. [16]

As opposed to weak anticipation, which he denotes "sophisticated guessing", strong anticipation is "a matter of (lawful) perceiving" — a two-term relation [17].

Direct perception does without the notion of representation and thereby provides a shortcut. Representational models, by contrast, have to deal with an additional instance, which needs to be compensated in order to avoid a delay. But, as Turvey's description of the walking creature shows, we often have to contend with both two- and three-term relations.

However, philosopher Inês Hipólito reminds us of the constraints enactivists set on representation, which exclude many forms of life (e.g. animals):

> Embodied and enactive cognitive science rejects the existence of representations or any form of model-like theorizing at any level that is not the level of a fully enculturated agent. (...) In short, for enactivists, representing requires enculturation and engagement with thinking and inference about a certain state of affairs in the world, i.e., they theorise. [18]

In the enactivist view, the construction and application of scientific models often suffer from a logical fallacy by confusing the map with the territory. While their models may well explain the behaviour of a system with the help of

> representational structures, such as symbols, information, or data, it does not follow that the system being investigated under the model possesses the properties of the model. [19]

Philosopher and psychologist Andy Clark belongs to the scientific community which recognizes the advantages of an embodied approach to perception and consciousness but stresses that we cannot do without representations in a model of ourselves and our environment. He proposes a model which

> ... will include various computational, representational and information-theoretic lenses that currently seem to provide our best understanding of the rich and complex space of adaptive trade-offs among neural, bodily, and environmental contributions and operations. [20]

Clark suggests a "predictive processing" (also denoted "predictive coding" or "hierarchical predictive coding") model of the brain, which dovetails with models of the embodied mind. It entails the action-in-perception approach with an action-oriented predictive brain [21]. In the predictive processing model, predictions meet sensory input:

> Bottom-up inputs are processed in the context of priors (beliefs/hypotheses) from layers higher up in the hierarchy. The unpredicted parts of the input (errors) travel up the hierarchy, leading to the adjustment of subsequent predictions, and the cycle continues. [22]

Clark provides a simple example of the bottom-up and top-down models: Returning to a steaming cup of coffee left in his office, he perceives the coffee through visual and olfactory signals (corners, edges, lines and colours). These combined shapes and relations then "activate bodies of stored knowledge" and lead to a perception of the coffee in its mug. This would be the bottom-up model of perception. The alternative, the top-down model, establishes the brain with expectations of the coffee-in-the-office scenario. Sensory signals (bottom-up signals) now meet top-down predictions about a probable scenario (a mug of coffee on a desk). If these predictions turn out to be true, we smoothly navigate towards the cup, while unexpected inputs are treated as new information [23]. This circular causal flow is described by Hipólito:

> [Predictive processing] aims to explain the brain's activity as constantly generating a model of the environment. The model is used to generate predictions of sensory input by comparing it to actual sensory input. (…) Bayesian formalism is used as a rule to update the model as new information becomes available. [24]

Predictive processing works by combining bottom-up sensory input with top-down probabilistic generative models. It is thus a model which includes representations. Predictive processing also endorses the embodied enactivist models, as the predictive brain is based on the enactive bodily interaction with the environment, in which perception is based on action.

To conclude, predictive coding is a model-based approach. It minimizes prediction error against the background of a more or less sophisticated model of the environment. And this model, in turn, mediates the organism's response to the environment [25]. The priors (expectations) in the predictive processing model make it an anticipatory system, as it involves a shortcut to perception and cognition.

For most purposes, the middle path between pure enactivist and representationist approaches is the most promising route to take when we look at anticipatory behaviour in humans and other living beings. Predictive processing with its two-term and three-term relations follows such a middle path. For inanimate systems, the two-term relation appears to be a more suitable candidate when we want to describe and predict their lawful behaviour. But, as we shall see in the following chapters, even the description of inanimate systems requires, in addition to initial conditions and natural laws, a setting of assignment conditions: Which conditions decide whether a system is part of a systemic whole or forms a separate inertial system? The answer to this question determines whether or not strong anticipation governs.

2.4 Obserpants, Environments and Systemic Wholes

When we talk about a system and its environment, we sometimes overlook the inbuilt dualism presupposed in many models. The idea that an organism adapts to its environment is generally accepted, but, as Turvey has pointed out, it entails a logical conundrum. Organism and environment entail different types of causality — a direct result of Newton's dualism, which assigns efficient cause to the environment and material cause to the system [26].

A brief reminder: The Aristotelian causal categories answer the question of why change occurs. The material cause of a house is concrete, wood or whatever was used to construct it. The formal cause of a house is its construction plan, its design. The efficient cause of a house is the agency which brings it forth, such as a bricklayer or carpenter. And the final cause of the house is its teleological purpose as envisaged by the home builder, e.g. to house a family or a business.

To Newton, the environment was simply everything that did not belong to the system under consideration. And this externality causes change in the system in the form of efficient causation — an effective forcing function. Rosen rejected this reduction of the environment into a mere forcing function. For complex systems, the organism–environment dualism is no longer tenable, as the Newtonian division into states and laws of motion cannot explain the behaviour of the living world, such as anticipation [27].

In the same vein, Turvey points out that the idea of an organism adapting to its environment suffers from the same logical predicament induced by the organism–environment dualism:

> Just as Newton's environment is known only through its effect on the system, the environment or niches of adaptation theory are known only through their organisms. [28]

The enactive approach described in the preceding section circumvents the dilemma which results from Newtonian dualism by treating a system and its environment as one systemic whole. Human beings both monitor and partake in their environment. From the neurosciences to the enactive paradigm to quantum mechanics, the participating role of the observer has been widely accepted. We cannot measure a system without interfering with it. Perception implies action which our environment constrains. Living beings and inanimate objects are always embedded in a context we cannot disregard when we describe their behaviour, so we need to see them not as isolated systems but as systemic wholes.

In order to avoid the somewhat unwieldy term observer-participant, I have coined the term "obserpant" [29], which I shall henceforth also use in this book. To what extent obserpants act as observers or participants determines their degree of involvement in their natural environment and social context [30].

Hence, our modelling relations should include the degree of involvement. This may be measured as the relation of those parts of the systemic whole which belong to the obserpant and those which belong to the system they interact with. Identifying the position of the interface between the obserpant and environment is a tricky matter, not only because the obserpant's sphere of influence is in constant flux but also because different obserpants would not always agree on where to set this interfacial cut.

If the environment is no longer reduced to an external forcing function, we may describe an organism and its environment as two coupled systems which, in turn, can be described as one systemic whole.

2.5 Weak and Strong Anticipation: Model-Based vs. Model-Free Architecture

Mathematician Daniel Dubois first coined the terms weak and strong anticipation. They differentiate between model-based and model-free architectures [31]. According to Dubois, weak anticipation is the property of a system to make predictions based on models, such as a model of the environment which predicts the weather. Carrying an umbrella because the weather forecast predicted rain is an example of someone making use of a forecasting model: The person's behaviour is determined by an expectation (the possibility of rain to come) based on a weather-forecasting model and is thus an example of weak anticipation. If, however, a baseball outfielder predicts the exact time and place at which he can intersect the ball's trajectory and catch it, his prediction is not based on an external model (it would take far too much time to calculate) and is therefore an example of strong anticipation. While weak anticipation requires an interpreting agent and is computationally demanding, strong anticipation results from natural interaction between the agent and environment.

While Rosen saw anticipation as a hallmark which distinguishes the living world from inanimate systems, Dubois argues that anticipation exists also in physical systems. He showed this, for instance, for electromagnetism and relativity transformations, where he deduced the relativistic corrective term of Newton's law of gravity from an anticipative effect [32].

Dubois introduced the notions of incursion (which stands for inclusive recursion) and hyperincursion — computational methods which calculate the present state of a system by taking into account its past and future states. In other words, the present state is thus determined by both initial and final conditions. To control and stabilize a system, Dubois showed that the insertion or removal of a delay can turn a chaotic system into a system with a stable attractor or, conversely, a stable system into a chaotic one. Against this background, he points out the fact that anticipation and delay are complementary concepts, i.e. that delay compensation is anticipation [33].

Dubois' concept of hyperincursion — an incursion with multiple solutions — also defines the present state as determined by both past and future states. However, in contrast to incursion, it is a range of multiple solutions to future states which determine the present [34]. Hyperincursion is a formalization of strong anticipation. However, as biological engineer

Damian Stephen and psychologist James Dixon point out, this does not imply that the causal order is reversed:

> At a first glance, hyperincursion might seem as though it flies in the face of a basic fundamental assumption of logical scientific thought, namely, that causes precede effects. However, hyperincursive logic need not entail that causes follow effects. Rather, it may be understood to express the multiple time scales on which anticipatory behavior unfolds. [35]

The researchers stress the fact that there is a wide range of phenomena which display long-range correlations on multiple scales, such as $1/f$ fractal scaling (see Chapter 5) in biology, meteorology, geology and astrophysics and interpret

> hyperincursive logic as expressing the concurrent effects of short-range behaviour and of long-range behaviour on the same fluctuation at time t in instances of these phenomena. Indeed, Dubois himself stressed the role of fractal scaling in hyperincursive logic. [36]

Computer scientist Mihai Nadin challenges Dubois' concepts of incursion and hyperincursion, claiming that they do not "satisfy the need to allow for a vector pointing from the future to the present" [37]. By contrast, psychologists Nigel Stepp and Michael Turvey support Dubois' notions:

> … delay-coupled systems themselves (…) do not contain any reference to future times (…). A solution to such systems, however, is $y(t) = x(t+\tau)$, which *does* contain a reference to the future. It is in this way that future states are implicit. [38]

Dubois' hyperincursion has been utilized in a variety of fields [39], including computation theory. Cellular automata, for instance, are based on discrete dynamics and are thus a fertile environment for hyperincursion. Strong anticipation manifests itself as the property of cells whose transition to the next state depends on potential future states. Mathematician and computer scientist Alexander Makarenko based his multi-valued solutions to cellular automata on Dubois' notion of hyperincursion [40].

Makarenko exemplifies the concept of multi-valuedness with Conway's Game of Life. Whereas the linear classical version of the game produces one state after another, the strong anticipative version generates multiple solutions with each step, thus creating a large number of branchings with multiple solutions:

> ...the main distinguishing feature of cellular automata with strong anticipation is the multiplicity of solutions, and multivaluedness of cell's states and of configurations at [the same] time ... [41]

While the classic version generated successive solutions, the strong anticipatory one produced simultaneous possible solutions with every successive step. In addition to more conventional applications, Makarenko suggests that the multi-valuedness, which results from introducing strong anticipation into cellular automata, may model Everett's interpretation of quantum mechanics. Multi-valuedness creates an ambiguity in the decision tree that leads to the emergence of a variety of solutions.

Stephen and Dixon adopt Dubois' differentiation between weak and strong anticipation [42]. While weak anticipation relies on a model which requires an interpreting agent, strong anticipation rules in a model-free architecture. It emerges from an embedded organism's actions and environmental constraints, i.e. from the natural behaviour of a coupled organism–environment system.

2.6 Two Types of Strong Anticipation

Whereas weak anticipation is always model-based, the concepts and theories of Noë, Turvey, Dubois (with his notion of hyperincursion), Stephen, Dixon and Stepp work without the internal model and are thus examples of strong anticipation. However, this notion requires a further distinction. Stephen and Dixon differentiate between local and global types of strong anticipation. The local type describes anticipatory synchronization, resulting from local coupling between organism and environment, as first described by Dubois [43]. The second type refers to a more global coordination between organism and environment [44].

2.6.1 *Local strong anticipation*

Stepp and Turvey adopt Dubois' differentiation between weak and strong anticipation and put it in a nutshell:

> Anticipation is weak if it arises from a model of the system via internal simulations. Anticipation is strong if it arises from the system itself via lawful regularities embedded in the system's ordinary behaviour. [45]

They point out the advantage of Dubois' notion of strong anticipation:

> The strategic importance of [strong anticipation] is that it invites a shift of focus from theorizing about a representation-anticipation relation to theorizing about a coupling-anticipation relation. (...) Rather than ask how the future is produced from an internal model, one asks about a lawful coupling (between organism and environment) that results in coordination with the future. [46]

To exemplify the difference between weak and strong anticipation, authors often draw analogies with sensory-motor adaptation to moving objects, such as tennis balls or baseballs [47]. In the so-called "outfielder scenario", the player is trying to catch a ball in flight. This may be done by using an external model which calculates the ball's trajectory. The point and time of landing are determined by physical laws which are modelled by the outfielder as he or she is trying to run towards the point of interception. If the outfielder relied on a model and had to calculate the curve of the trajectory, the resulting behaviour would be described by Stepp and Turvey as weak anticipation. The researchers make a significant differentiation: The weak anticipation scenario deals with two separate systems consisting of the outfielder and the ball-in-flight, where the ball-in-flight is the outfielder's environment. A weakly anticipatory action would consist of the outfielder calculating when and where the ball will land and then running to that point, which would make him or her an external observer of the action.

However, this is not what happens in real life, not just because it would take far too long but because it is unnecessary. The outfielder will most likely keep his or her eyes on the ball and readjust motion and posture when closing in on the point of interception. The ball and player become one systemic whole. Stepp and Turvey conclude the following:

> … by causing the agent to become a part of the system that the agent is anticipating, the non-analytic method takes advantage of existing information about the future. [48]

Strong anticipation regards the outfielder and the ball-in-flight as one single system, i.e. a systemic whole which displays its own dynamics.

For Stepp and Turvey, strong anticipation basically means anticipatory synchronization, as occurs in coupled systems such as a human being locking into the rhythms of his or her environment [49].

The authors list five necessary conditions for strong anticipation: Strong anticipation is an achievement of a systemic whole. Rather than dealing with at least two separate systems as in weak anticipation (where an agent with an internal model of his or her environment predicts the future of that environment), strong anticipation is a property of the systemic whole consisting of agent and environment. In addition, the subsystems of the systemic whole must be coupled in such a way that they can follow the lawful regularity of the systemic whole. Furthermore, the systems involved in anticipatory behaviour do what they naturally do — the ball flies and the outfielder runs, watching the ball. The systems involved do not need to be aware of the role they play in the emergence of anticipatory behaviour or the fact that strong anticipation manifests itself — they just do what they do. Another hallmark and necessary condition of strong anticipation is the fact that, at some level of description, an anticipatory system is purely reactive. If one regards one subsystem, e.g. the flying ball, in isolation, it is purely reactive as it follows Newton's Law of Gravitation. Only in conjunction with at least one other system does it display strong anticipation if regarded as one systemic whole. Finally, anticipatory systems implicitly relate to future states (as opposed to weakly anticipatory systems, which explicitly refer to the future):

> Strong anticipation (…) does not explicitly concern itself with the future. Strongly anticipatory systems go about their normal functioning,

and are implicitly affected by the future because of how they are put together. [50]

2.6.2 *Global strong anticipation*

In 2008, Kelty-Stephen, Stepp, Dixon and Turvey first showed that synchronization with a chaotic signal could not be based on local cues but on global coordination between the organism and environment [51]. They conducted an experiment in which subjects tried to synchronize their tapping with a chaotic, i.e. unpredictable, signal in the form of a metronome. The researchers found that there was little coupling in the short term but in the long term, participants' dynamics coordinated with those of the environment.

In a reanalysis of their 2008 paper, Stephen and Dixon point out that the intertap intervals were sensitive to coordination on multiple time scales [52]. They suggested that strong anticipation is a multi-level phenomenon which can be understood as the coordination of fluctuations between levels. This means that the long-term correlations within an organism adapt to the long-term correlations in environmental structures (and, possibly, vice versa). However, the nature of the long-term synchronization was of a statistical nature (more on the precise nature of this global synchronization in Chapter 5, when we look at the concept of fractal time).

Stephen and Dixon stress that it is important to keep in mind that weak anticipation and both types of strong anticipation are based on very different casual assumptions. While weak anticipation is explained by internal models, local strong anticipation results from the coupling of the organism and environment. And global strong anticipation, according to Stephen and Dixon, can be explained by the coordination between the multi-level structures both in perceptual-motor fluctuations and in environmental fluctuations. They leave open the question as to whether local coupling and global coordination are causally interlinked [53].

Whether our approaches are simulation- and model-based or do not require the concept of an internal model or representations, the problem of the obserpant's perspective remains to be dealt with. The following section looks at the notions of endo- and exo-perspectives and the interface between the obserpant and the rest of the world.

2.7 Our Models Are Anthropocentric

Before we set out to investigate manifestations of (and precursors to) strong anticipation, a reminder of implicit and often hidden constraints in our descriptions of the world is necessary. Our physical theories are secondary constructs of our primary experience of time, space and causality [54].

Psychologist Ernst Pöppel stresses that we have to take into account *a priori* the processing in our brains when we consider our theories about ourselves and the world. An example is our experience of time, which manifests itself as succession, simultaneity, duration and the Now. Our brains make up for the temporal incongruence of audio, visual or tactile perceptions through postdiction by integrating them into a gestalt. What is perceived as successive and what as simultaneous depends not only on the different speeds at which the signals are propagated in our environment but also on the differing processing times of the various modalities (visual, auditory and tactile) in our brains. Pöppel defined a simultaneity horizon at a radius of 10–12 meters from the observer, at which audio and visual signals are perceived as being simultaneous. The experience of duration and the extension of our Now also depend on the brain's capacity to integrate perceptions and generate perceptional gestalts [55]. (More on the *a priori* and the structure of time in Chapter 5.) Pöppel claims that our primary experiences of time precede any physical or semantic notions of time and act as constraints on our models of the world:

> (…) the search for the conditions of rendering possible any experience of time in the real world is determined by mechanisms of our brains, which condition our experience of time. It is not possible to conceive of a theoretical concept of time in physics (e.g. Newton's 'absolute' time), which pretends to exceed experience of time. Therefore, I suggest to regard the physical concepts of time (be it the Newtonian one, that of Einstein or Prigogine) as a secondary construct derived from our primary experience of time. [56]

Another constraint to creating an ideal model is the fact that our perspective is distorted as a result of the microscopic movements in our bodies, including our brains, which together form an interface between ourselves and our environment. This interface is a constraint and our only access to the world: What we see is not the world but an interfacial cut. Biochemist and chaos theorist Otto Rössler, the founder of the science of the interface ("interfaciology"), stresses that the decision which microscopic movements of particles are assigned to the observer and which to the rest of the world determines the state of our interface and thus shapes our interfacial distortions. These distortions of time and space manifest themselves as our "Now" [57].

Rössler uses the term "endo-perspective" for the view from within an observer and "exo-perspective" for a view from an outside vantage point [58]. The endo-obserpant is not in a position to determine the dividing line between sensory-motor or cognitive processes which are embodied and those which have been outsourced. The point is that, from the perspective of an external obserpant, those outsourced elements and structures belong to the environment. An example is a walker who has embodied a walking stick, which is clearly visible from his smooth movements and the fact that he or she is oblivious to the fact that an obserpant extension has been embodied [59]. The extended interface between the hand and stick has become transparent, the walker's hand feels the ground through the stick in his or her hand. The walker and stick have become one systemic whole. From the outside, however, two systems are visible, namely the walker and the stick, and although the external obserpant could deduce from the smooth movements that, for the walker, the stick has become an extra limb, he or she can still clearly set the interfacial cut between the hand and stick (rather than between the new systemic whole and its environment).

Rössler equated his exo-observer with Laplace's demon, who himself is not part of the world he observes and can describe and predict the past and future of every particle in the universe, given their initial conditions and the natural laws governing them. Such an observer would be a sheer observer, as he could not participate in or influence the fate of the universe, although he would be fully aware of the position and momentum of every single particle. By contrast, endo-observers would be both part of the universe they observe and also participate through their interactions.

However, endo-observers are barred from sheer participation, i.e. the idea of becoming one with the universe, as they perceive the rest of the world through an interface and are therefore also capable of observation.

We are confined to our endo-perspectives and the exo-perspective remains an unattainable idealization. A positive side effect of a universe of endo-perspectives attempting to communicate is the fact that there is no privileged observer frame — it is a truly emancipating scenario, which assigns the same level of "truth" to each endo-perspective [60].

Astrophysicist Rosolino Buccheri and musicologist Marina Alfano endorse the endophysical approach and suggest giving it a central role in our scientific endeavours by placing the interacting obserpant into the centre of our most basic considerations. This can be achieved by embedding conventional scientific rationality into a wider context which recognizes that the obserpant co-evolved with his or her environment. The development of human cognition started with physical interaction, co-oscillation and resonance with the environment. This evolutionary inheritance still reverberates in our interactions with our environment and provides a wider framework to host more traditional scientific approaches [61].

Neither sheer observers nor sheer participants can communicate or take part in the intersubjective construction of a scientific paradigm. They do not belong to the realm of science but to that of faith. Science starts in-between with the obserpant. As endo-obserpants, we are neither sheer participants nor sheer observers but somewhere between these two extremes.

The construction of models by obserpants is constrained by their respective perspectives and physical makeup. This idea that obserpants should be treated as cybernetic systems in their own right has been taken into account by 2nd-order cybernetics. And as the oberpants' ontologies of the 2nd-order domain need to be contextualized within a common framework, 3rd-order cybernetics achieves this by taking account of both the system–context interface and the observer–involvement interface [62]. Thus, models and scientific theories are constructed by embedded steersmen.

A detailed account of the increasing weight of the subjective role of the obserpant in the construction of models in history and the accompanying awareness of their anthropocentric nature exceeds the scope of this book. To name but a few of the most important game changers: The Newtonian paradigm did not allow for observer frames. These were first introduced by Albert Einstein in his Special Theory of Relativity, which assigned the same "truth" to conflicting observations from different inertial systems. The same is true for Rössler's endophysics, which shows that there are no privileged observer perspectives (if one disregards Laplace's demon). The notion of the observer-participant first appeared in quantum mechanics, with the observer-participant collapsing the wave function as a result of his or her measurement, thus generating reality through interference. And the advent of psychoanalysis positioned the individual at the mercy of unconscious drives, which determine his or her action, although they remain elusive.

From a personal endo-perspective, all the obserpant can do is construct a model of the world which is waterproof and falsifiable [63]. It is only when our expectations are not met that we have to acknowledge the possibility of a force whose functions are not part of our neural processes. Rössler refers to this method of waterproofing as intra-dream consistency [64]. As long as our perceptions of the world are not contradictory, our world is intact and any external functionalities which differ from those of our neural processes will not be recognized as such. Contradictions are brought to our attention when the assumptions upon which our expectations about the world have emerged prove to be false [65].

It is true that much of the interaction which occurs both within the obserpant as well as between the obserpant and his or her environment happens unconsciously. However, we are given — now and then — hints which reveal hidden assumptions about the way we construct the world. Merleau-Ponty revealed one such blind spot, namely that the position of our body in space occupies, as well as our perceptual range, determine how we perceive the world and relate to it (see Chapter 10). Our visual field, for instance, is limited and directed: we see but one side of a building but assume that it consists of more than the front without walking around to verify our assumption. We are not simply embedded in our environment — we are always part of what we observe, albeit as a mere constraint.

Taking a closer look at strong anticipation may reveal some of our hidden assumptions about the world. This book sets out to reveal some of them.

References

[1] A. Rosenblueth and N. Wiener, The role of models in science. *Philosophy of Science* 12, 316, 1945.
[2] R. Rosen, *Anticipatory Systems: Philosophical, Mathematical and Methodological Foundations*, 2nd edition, Springer, 2012, p. 7 (First published by Pergamon Press 1985).
[3] *ibid*, p. 7.
[4] *ibid*, p. 8.
[5] *ibid*, pp. 71–72.
[6] *ibid*, p. 400.
[7] *ibid*; Rosen's mathematical approach was based on category theory, a general theory of structures and their relations, which shows various ways in which different types of mathematical structures are related to each other. For a general introduction, I recommend I. David, *Spivak's Category Theory for the Sciences*, MIT Press, 2014.
[8] A. Noë, *Out of Our Heads*, Hill and Wang, New York, 2009, p. 1.
[9] *ibid*, p. 164.
[10] *ibid*, p. 108.
[11] *ibid*, p. 211.
[12] I. Hipólito, Cognition without neural representation: Dynamics of a complex system. *Frontiers in Psychology Section on Theoretical and Philosophical Psychology*, 12 January 2022.
[13] M. T. Turvey, *Lectures on Perception. An Ecological Perspective*, Routledge, New York, 2019, pp. 27–28.
[14] *ibid*, p. 29.
[15] *ibid*, p. 29.
[16] *ibid*, p. 416.
[17] *ibid*, p. 415.
[18] I. Hipólito, Cognition without neural representation: Dynamics of a complex system. *Frontiers in Psychology*, 12, 8, 2022. Art. 643276.
[19] *ibid*, pp. 8–9.
[20] A. Clark, *Supersizing the Mind. Embodiment, Action and Cognitive Extension,* Oxford University Press, New York, 2008, p. 217.

[21] A. Clark, *Surfing Uncertainty. Prediction, Action and the Embodied Mind*, Oxford University Press, 2016/2019, p. 295.

[22] *ibid*, p. 30.

[23] *ibid*, pp. 13–14.

[24] I. Hipólito, Cognition without neural representation: Dynamics of a complex system. *Frontiers in Psychology*, 12, 7, 2022. Art. 643276.

[25] P. Calvo, F. Baluška and A. Sims, "Feature detection" vs. "predictive coding" models of plant behavior. *Frontiers in Psychology*, 7, 1505, 4, 2016.

[26] M. T. Turvey, *Lectures on Perception. An Ecological Perspective,* Routledge, New York, 2019, p. 17ff.

[27] R. Rosen, *Anticipatory Systems: Philosophical, Mathematical and Methodological Foundations*, 2nd edition, Springer, New York, 2012, p. 7 (First published with Pergamon Press, Oxford, 1985).

[28] M. T. Turvey, *Lectures on Perception. An Ecological Perspective*, Routledge, New York, 2019, p. 19.

[29] S. Vrobel, Analogue and digital bases of trust. Talk given at *InterSymp*, Baden-Baden, Germany, August 2014, and S. Vrobel, A new kind of relativity: Compensated delays as phenomenal blind spots. *Progress in Biophysics & Molecular Biology* (Elsevier, London), 119, 303–312, 2015; Please note that I have substituted other researchers' usage of "observer" or "observer-participant" in this book wherever it is conducive to clarity. This does, of course, not imply that others have used the term "obserpant".

[30] O. van Nieuwenhuijze, The equation of health. In: D. M. Dubois (ed.) *International Journal of Computing Anticipatory Systems* (CHAOS, Liège), 22, 235–249, 2008.

[31] D. M. Dubois, Mathematical foundations of discrete and functional systems with strong and weak anticipations. In: M. V. Butz, O. Sigaud and P. Gérard (eds.) *Anticipatory Behavior in Adaptive Learning Systems.* Lecture Notes in Computer Science, Vol. 2684, Springer, Berlin, 2003. The concepts were mentioned already in 1998, when Dubois denoted the concepts of strong and weak anticipation as internalist and externalist. In: D. M. Dubois, Introduction to computing anticipatory systems, *International Journal of Computing Anticipatory Systems* (CHAOS, Liège, Belgium), 2, 3–14, 1998.

[32] D. M. Dubois, Review of incursive, hyperincursive and anticipatory systems — foundation of anticipation in electromagnetism. In D. M. Dubois (ed.) *Computing Anticipatory Systems: CASYS'99, AIP Conference Proceedings*, Vol. 517, The American Institute of Physics, 2000, pp. 3–30; D. M. Dubois, Anticipative effect in relativistic physical systems,

exemplified by the perihelion of the planet mercury. In: R. L. Amoroso, B. Lehnert and J.-P. Vigier (eds.) *Beyond the Standard Model: Searching for Unity in Physics, Proceedings of the Paris Symposium Honoring the 83rd Birthday of Jean-Pierre Vigier*, published by The Noetic Press, Orinda, USA, 2005, pp. 119–132.

[33] D. M. Dubois, Mathematical foundations of discrete and functional systems with strong and weak anticipations. In: M. V. Butz, O. Sigaud and P. Gérard (eds.) *Anticipatory Behavior in Adaptive Learning Systems.* Lecture Notes in Computer Science, Vol. 2684, Springer, Berlin, 2003, p. 131 and personal communication, 2006.

[34] "An anticipatory system is a system which contains a model of itself and/or of its environment in view of computing its present state as a function of the prediction of the model. With the concepts of incursion and hyperincursion, anticipatory discrete systems can be modelled, simulated and controlled. By definition an incursion, an inclusive or implicit recursion, can be written as: $x(t+1) = F [..., x(t-1), x(t), x(t+1), ...]$ where the value of a variable $x(t+1)$ at time $t+1$ is a function of this variable at past, present and future times. This is an extension of recursion. Hyperincursion is an incursion with multiple solutions. For example, chaos in the Pearl-Verhulst map model $x(t+1) = a.x(t).[1 - x(t)]$ is controlled by the following anticipatory incursive model $x(t+1) = a.x(t).[1 - x(t+1)]$ which corresponds to the differential anticipatory equation $dx(t)/dt = a.x(t).[1 - x(t+1)] - x(t)$." (D. M. Dubois, Computing anticipatory systems with incursion and hyperincursion. In: *Computing Anticipatory Systems: CASYS — First International Conference, AIP Conference Proceedings*, Vol. 437, The American Institute of Physics, 1998, p. 3.)

[35] D. G. Stephen and J. A. Dixon, Multifractal cascade dynamics modulate scaling in synchronization behaviours. *Chaos, Solitons & Fractals* (Elsevier), 44(1–3), 161–168, 2011.

[36] *ibid*, p. 161.

[37] M. Nadin in R. Rosen: *Anticipatory Systems: Philosophical, Mathematical and Methodological Foundations.* Second Edition, Springer, New York, 2012, p. xli

[38] N. Stepp and M. T. Turvey, On strong anticipation. *Cognitive Systems Research*, 11(2), 158, 1 June 2010.

[39] Against the background of their notion of collective beings, mathematician Gianfranco Minati and physicist and cognitive scientist Eliano Pessa also use Dubois' equations for a simple formal formulation of an anticipatory system: "Formally, an anticipative system is a system X whose dynamical

evolution is governed by the equation. X(t+1) = F(X(t), X*(t+1)) where X*(t+1) is X's anticipation of what its state will be at the time (t+1)." (G. Minati and E. Pessa, *Collective Beings*, Springer, New York, 2010, p. 21.

[40] D. M. Dubois, Generation of fractals from incursive automata, digital diffusion and wave equation systems. *BioSystems*, 43, 97–114, 1997; D. M. Dubois, Incursive and hyperincursive systems, fractal machine and anticipatory science. *AIP Conference Proceedings*, Vol. 573, The American Institute of Physics, 2001, pp. 437–451; A. Makarenko, Multivaluedness in cellular automata with strong anticipation and some prospects for computation theory. *WSEAS Transactions on Information Science and Applications*, 17, 69–78, 2020.

[41] *ibid*, p. 72.

[42] D. G. Stephen and J. A. Dixon, Multifractal cascade dynamics modulate scaling in synchronization behaviours. *Chaos, Solitons & Fractals* (Elsevier), 44(1–3), 160–168, 2011.

[43] D. M. Dubois and G. Resconi, Introduction to hyperincursion with applications to computer science, quantum mechanics and fractal processes, *CC-AI*, Vol. 10, NOS 1–2, 1993, pp. 109–148; D. M. Dubois, Theory of incursive synchronization and application to the anticipation of a chaotic epidemic, *International Journal of Computing Anticipatory Systems*, 10, 3–18, 2001; D. M. Dubois, Theory of incursive synchronization and application to the anticipation of delayed linear and nonlinear systems. In: *Computing Anticipatory Systems: CASYS 2001, AIP Conference Proceedings*, Vol. 627, The American Institute of Physics, 2002, pp. 182–195.

[44] D. G. Stephen and J. A. Dixon, Multifractal cascade dynamics modulate scaling in synchronization behaviours. *Chaos, Solitons & Fractals* (Elsevier), 44(1–3), 160–168, 2011.

[45] N. Stepp and M. T. Turvey, On strong anticipation. *Cognitive Systems Research*, 11(2), 148, 1 June 2010.

[46] M. T. Turvey, *Lectures on Perception. An Ecological Perspective*, Routledge, New York, 2019, p. 413.

[47] N. Stepp and M. T. Turvey, On strong anticipation. *Cognitive Systems Research*, 11(2), 1 June 2010.

[48] *ibid*, p. 152.

[49] It should be noted that the terms rhythm and oscillation have the same meaning and my usage of these terms depends on the context. For human gait or music, for instance, the term 'rhythm' is more suitable than 'oscillation'.

[50] N. Stepp and M. T. Turvey, On strong anticipation. *Cognitive Systems Research,* 11(2), 137–138, 1 June 2010.

[51] D. G. Kelty-Stephen *et al.*, Strong anticipation: Sensitivity to long-range correlations in synchronization behaviour. *Physica A,* September 2008.

[52] D. G. Stephen, and J. A. Dixon, Multifractal cascade dynamics modulate scaling in synchronization behaviours. *Chaos, Solitons & Fractals* (Elsevier), 44(1–3), 160–168, 2011.

[53] *ibid.*

[54] Linguist George Lakoff and cognitive scientist Rafael E. Núñez claim that mathematics, too, is grounded in the human cognitive apparatus, and is thus a secondary construct. (G. Lakoff and R. E. Núñez, *Where Mathematics Come from: How the Embodied Mind Brings Mathematics into Being*, Basic Books, 2001.)

[55] E. Pöppel, Erlebte Zeit und Zeit überhaupt: Ein Versuch der Integration. In: H. Gumin, and H. Meier (eds.) *Die Zeit. Dauer und Augenblick*, Piper, Munich, 1989; E. Pöppel, *Grenzen des Bewußtseins — Wie kommen wir zur Zeit, und wie entsteht die Wirklichkeit?* Insel Taschenbuch, Frankfurt, 2000, pp. 38–42.

[56] E. Pöppel, Erlebte Zeit und Zeit überhaupt: Ein Versuch der Integration. In: H. Gumin, and H. Meier (eds.) *Die Zeit. Dauer und Augenblick*, Piper, Munich, 1989, p. 380 (my translation).

[57] O. E. Rössler, *Endophysics*. World Scientific, Singapore, 1998.

[58] The term 'endophysics' was suggested by David Finkelstein in a correspondence with Rössler. In: O. E. Rössler, *Endophysics,* World Scientific, Singapore, 1998; Drawing on Rössler's concepts, Dubois coined the terms "endo- and exo-anticipation" and assigned them to strong and weak anticipation, respectively, a differentiation he had defined earlier. (D. M. Dubois, Mathematical foundations of discrete and functional systems with strong and weak anticipations. In: M. V. Butz, O. Sigaud and P. Gérard, (eds.) *Anticipatory Behavior in Adaptive Learning Systems.* Lecture Notes in Computer Science, Vol. 2684, Springer, Berlin, 2003.

[59] A. Clark, *Supersizing the Mind. Embodiment, Action and Cognitive Extension*, Oxford University Press, New York, 2008, pp. 30–37.

[60] O. E. Rössler, *Endophysics*, World Scientific, Singapore, 1998.

[61] M. Alfano and R. Buccheri, Personal communication 2012 and M. Alfano and R. Buccheri, What kind of time for a time machine? *RPJ Web of Conferences*, 58, 04001, 2013.

[62] O. van Nieuwenhuijze, Integral Health Care. Dissertation draft, 1998 (unpublished); O. van Nieuwenhuijze, The equation of health. In: D. M. Dubois (ed.) *International Journal of Computing Anticipatory Systems* (CHAOS, Liège), 22, 235–249, 2008 and personal communication, 2005.

[63] K. R. Popper, *The Myth of the Framework.* In: M. A. Notturno (ed.), Routledge, New York, 1994.

[64] O. E. Rössler, *Descartes' Traum. Von der unendlichen Macht des Außenstehens.* Audio-CD, Supposé, Cologne, 2002.

[65] K. R. Popper, *The Myth of the Framework.* In: M. A. Notturno (ed.), Routledge, New York, 1994.

Chapter 3

Extended and Reduced Observant Perspectives: Boundary Shifts

So far, we have looked at weak and strong anticipation as defined by Dubois, Stepp and Turvey, and Stephen and Dixon. We differentiated between model-based and model-free anticipatory architecture and between local and global strong anticipation. Local strong anticipation can manifest itself as anticipatory synchronization, as defined by the conditions laid out by Stepp and Turvey. Global strong anticipation unfolds on multiple scales and manifests itself as the coordination of both the organism's and the environment's long-term correlations of fluctuations. Examples of local and global strong anticipation are presented in Chapter 4.

This chapter deals with a specific type of local strong anticipation, namely obserpant extensions which result in spatial and temporal boundary shifts. These shifts occur both to the inside and the outside and facilitate or constrain strong anticipative behaviour. Particular emphasis is put on the formation of systemic wholes as a result of boundary shifts towards the outside, which extend obserpants' spatio-temporal sphere of influence. What was part of the outside world has become part of the obserpant. In this process, the boundary shift and the interface between the inside and outside remain transparent to the obserpant. While a sense of agency is retained, this boundary turns out to be highly negotiable.

3.1 Boundary Shifts: Spatially Extended Observants as Examples of Local Strong Anticipation

Obserpant extensions can be external add-ons, such as clothes and other wearables, cars, weapons, smartphones, walking sticks, glasses, hearing aids or more recent applications like hearing wristlets. Then there are additions which were previously external and are now inside the obserpant's body: hip replacements, cardiac pacemakers, neural implants, etc. More sophisticated extensions like linguistic and semiotic skills as well as specific social and cultural conventions are virtual extensions.

Matthew Botvinick and Jonathan Cohen's rubber hand illusion is a classic example of an obserpant addition. It functions as an invisible bridge which closes a spatio-temporal gap that existed between the obserpant and the rubber hand when no sense of agency had yet emerged and the rubber hand was still part of the environment.

The original experiment showed how bodily self-attribution emerges in a multi-modal interaction, as a result of an intersensory bias illusion [1]. Subjects were placed at a table, with their left arm on the table surface. Next to it, a screen was placed so as to hide the arm from the subjects' view. A rubber model of an arm, including a hand, was placed on the table in front of the subjects. They were asked to gaze at the rubber hand while both the artificial and the subject's hidden hand were simultaneously stroked with a brush for 10 minutes. Subjects reported that, after a while, they felt the brush on the rubber hand rather than on their real left hand.

The intermodal matching (vision, touch and proprioception) resulted in the self-attribution of the rubber hand. It became part of the obserpant when the boundary between self and nonself (the rest of the world) was shifted outwards. The interfacial shift assigned what used to be part of the environment, i.e. the rubber hand, to the obserpant. The subject and the rubber hand now formed one systemic whole, composed of the obserpant and (incorporated) environment, whereby the systemic whole extended the subject's proprioceptive space. And as the compensation of the spatio-temporal gap, which accompanied the self-attribution of the rubber hand, was transparent to the subjects, they were unaware of the compensatory process which guided their perception.

A more familiar example of a spatial observant extension is the walking stick which, if used frequently, acquires the smoothness of motion of an extra or extended limb [2]. Like wearables, it is an interfacial extension between our body and the rest of the world. The interfacial cut between the walker and environment is shifted outwards and is now located between the tip of the walking stick and the environment.

When the walker is no longer aware of the interfaces involved — the first where the observant's hand grasps the handle and the second where the stick's tip touches the ground — the observant and the walking stick merge into a new systemic whole. What was once a tool for walking has become an extension of the observant's arm.

Since the interfaces are transparent to the walker, the spatio-temporal distance and delay between the walker without the stick embedded in an environment and the walker with the stick embedded in an environment have been compensated. But what is transparent to the walker is visible from an external perspective. However smooth the gait, an onlooker would not regard the walker and the stick as a systemic whole. The observant's transparent interfaces are visible from an external perspective only. As in the rubber hand illusion, the compensation of the spatio-temporal gap is transparent to the observant.

Following Dubois' definition that anticipation and delay are complementary concepts (which means that delay compensation is anticipation), I shall suggest the following four conditions to qualify as local strong anticipative systems for extended observants:

(1) A boundary shift towards the outside occurs.
(2) The boundary shift results in the formation of a new systemic whole.
(3) The interfaces which have been merged or have newly emerged are transparent to the observant.
(4) The observant retains a sense of agency after the boundary shift and spatio-temporal compensation.

Conditions (1–3) are based on Stepp and Turvey's requirements for strong anticipation (see Section 2.6.1). Condition (4) is a requirement for extended observants only. Note that my four conditions for strong anticipation do not exclude representations, which are strictly assigned to weak anticipation by Stepp and Turvey, as well as Stephen, Dixon and Stepp. The need to include representations in the creation of new systemic wholes and the possibility to assign a sense of agency will become clear in the following sections and chapters.

In the example of the walker and stick creating a new systemic whole, one may argue that the two components are not tightly coupled, as the walking stick is passive. From an enactivist point of view, however, the walking stick offers resistance and constrains the walker's motion and posture. Still, this does not entail that the walker and stick are interdependent.

By contrast, wearers of sophisticated exoskeletons are coupled to an active environmental device which gives the walker active feedback. *AnDy* is a European research project with the aim of creating software and hardware which allow robots to "fully describe and predict the whole-body dynamics of the interaction between human and robots." [3]

To achieve this, the wearable *AnDySuit* tracks and stores human motions and provides the extensive data pool for the *AnDyModel*. *AnDyModel* generates ergonomic and anticipatory models of human motion and dynamics on the basis of these data. *AnDyControl*, an anticipative controller, uses these models to assist humans by predicting their movements. The probabilistic model draws on a pool of data, compares the agent's motion with that data and estimates the most likely next step. Although there is genuine interaction between the two components, the exoskeleton is a feedforward controller and the walker tries to maintain balance as a feedback controller.

Serena Ivaldi, Research Scientist at *AnDy*, points out that one of the metrics which will be crucial for collaboration is anticipation. This is true both for cobots (co-operative robots) and exoskeletons:

> Anticipation is related to the amount of synchronization needed to successfully complete the collaborative task, and it is strongly related to the prediction capabilities of the robot. In a situation of minimal anticipation, the robot does not need to anticipate the human partner's positions or forces to perform the collaborative task. If the robot is capable of classifying and recognizing the current action performed by the human, it can choose appropriate control settings to adapt to each action/task.

> If it can recognize the goal of a reaching movement, or the intent of motion, it can proactively provide a support control action. If it is able to predict the positions and forces that the human partner will do, it can use this information to choose its future control actions and postures as well. [4]

The ideal exoskeleton displays anticipative behaviour as the coupling allows the robotic device to make predictions about the possible future behaviour of the wearer. However, at first, the interfacial shift is only partly consummated. Although the joint movement of the exoskeleton and its wearer is achieved by the system as a whole, i.e. the exoskeleton is a physical extension of the walker's body, the wearer is at first still aware of the coupling. Transparency of the interfacial cuts would emerge only after full adaptation to the robotic device. Then the wearer retains a sense of agency when he or she walks with the exoskeleton, which suffices to classify it as an example of local strong anticipation for extended obserpants.

Ivaldi notes that the most important prerequisite for the collaboration of cobots and other devices such as exoskeletons is anticipation. She observed that there is a strong correlation between the degree of mutual care and anticipative faculties, which manifests itself in the degree of synchronization between the robotic device and the human being. Figure 3.1 depicts the envisaged development of robots towards strong anticipative devices.

Although an exoskeleton's actions are based on a probabilistic model, it fulfils, as a systemic whole together with the obserpant, the four conditions for strong anticipation in extended obserpants.

3.2 Boundary Shifts: Spatially Reduced Obserpants

Spatial obserpant extensions result in altered proprioception through the formation of a new systemic whole and transparent interfaces. The interfacial cut between the obserpant and environment is shifted outwards.

The opposite case, spatially reduced obserpants, also results in altered proprioception. The interfacial cut, however, is shifted inwards. And, rather than being anticipative, the shift results in confusion and surprise and sometimes in an illusory reversal of causal order. As with obserpant

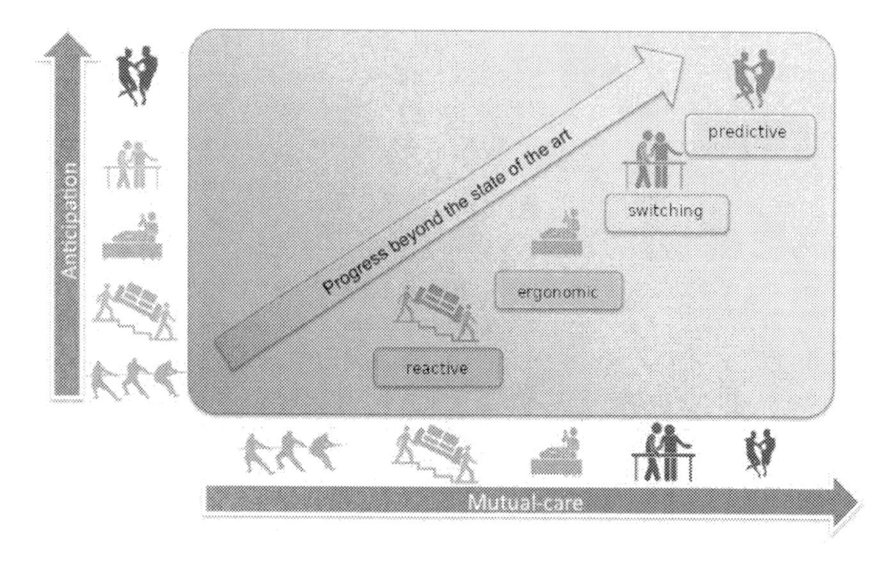

Fig. 3.1. A sketch of the collaborative control objectives, with respect to the different levels of mutual care and anticipation, used as metrics for collaboration: Progressing types of collaboration (reactive, ergonomic, switching and predictive) [5].

extensions, obserpant reductions cause a recalibration of proprioception because his or her hypothesis about the world is not confirmed.

An example is obserpants who assign parts of their bodies to their environment and no longer recognize them as their own. Such states result from neurological disorders, strokes or accidents. Neurologist Oliver Sacks reports the fate of a man who woke up one morning and found a strange leg in his bed. He was admitted to hospital for nocturnal motor disorders but was otherwise fine. When Sacks was called in, the man had fallen out of bed and was staring at his left leg. He seemed confused and refused to return to bed, telling Sacks that he was fine until he awoke from his nap and found the strange leg in his bed. The man tried to throw it out, only to find himself on the floor next to the bed and the alien leg seemed to have attached itself to his body. He was utterly surprised and confused and started pounding the leg. When Sacks advised him to stop hurting himself, the man insisted that his left leg was not his own — a conviction that did not change when the doctors pointed out to him that it was firmly connected to his pelvis. When asked about the whereabouts of his own left leg, he declared that it was missing and that he had no idea where it was.

His leg, which the patient claimed no longer belonged to him, was now assigned to the environment. The interfacial cut was shifted inwards [6].

Another example of a reduced obserpant is the patient suffering from unilateral neglect, a condition where the obserpant ignores part of his or her egocentrically encoded space [7]. Such a person may be completely unaware of what is happening, for instance, on the left side of their visual field and shave only the right part of their face or eat only the food which is placed on the right side of their plate. When asked to draw a clock, they only draw one half. The left hemisphere is not merely invisible to such individuals; it virtually does not exist for them.

These patients are examples of spatial boundary shifts towards the inside — the obserpants are deprived of part of their body or half of their peripersonal space. In unilateral neglect, the left hemisphere exists only for other onlookers but not for the patients. In the case of the patient not recognizing his own leg, the ownership of the leg was indisputable for bystanders. When a boundary shift occurs, internal and external obserpants are strictly separate systems. Their differing perspectives are made formally comparable in Chapters 9 and 10.

In the case of neglect, the "lost" environment can be regained by merging it with the patient into one systemic whole. This can be achieved by guiding the patient into "growing" an extra limb. Shifting the interfacial cut between the obserpant and environment has proved to be a promising treatment. It is a compensatory act which meets all four requirements for local strong anticipation in extended obserpants. Section 3.3 looks at how phantom limb patients "re-connect" the missing limb by virtually extending their body map with the help of a mirror box and how tools for reaching extend the obserpant's personal space.

3.3 Boundary Shifts: Recapturing Spatial Obserpant Reductions as Examples of Local Strong Anticipation

Intermodal matching, which brought about the rubber hand illusion, has proven a useful tool for phantom limb patients, who suffer from pain in the amputated limbs. Neuroscientist Vilayanur Ramachandran re-connected such patients with their no longer existing limbs by

Fig. 3.2. Dr. Vilayanur S. Ramachandran and psychology student Matthew Marradi with the original mirror box [8].

virtually extending their body map with the help of a mirror box (see Fig. 3.2)

In his seminal book *Phantoms in the Brain*, Ramachandran describes the device he invented:

> The box is made by placing a vertical mirror inside a cardboard box with its lid removed. The front of the box has two holes in it, through which the patient inserts her 'good' (say, the right one) hand and her phantom hand (say, the left one). (…) The patient is then asked to view the reflection of her normal hand in the mirror and to move it around slightly until the reflection appears to be superimposed on the felt position of her phantom hand. She has thus created the illusion of observing two hands,

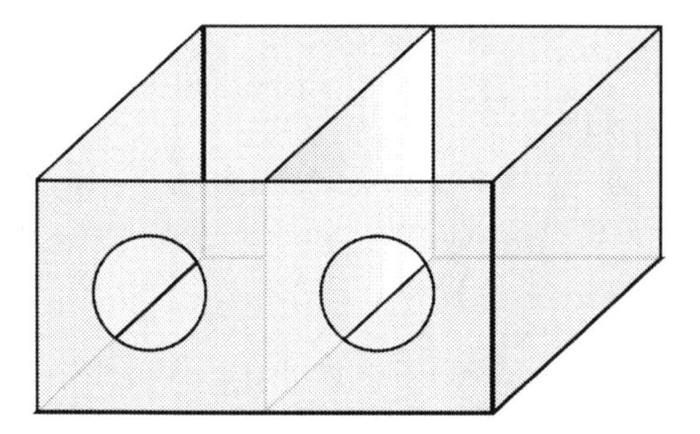

Fig. 3.3. Diagram of a mirror box [11].

> when she is only seeing the mirror reflection of her intact hand. If she now sends motor commands to both arms to make mirror symmetric movements, as if she were conducting an orchestra or clapping, she of course 'sees' her phantom moving as well. Her brain receives confirming visual feedback that the phantom is moving correctly in response to her command. [9]

Ramachandran's phantom pain patient who suffered from pain in his amputated arm was asked to insert his existing arm into a mirror box (see Fig. 3.3), which reflected the arm. If he then looked at the mirror box from a certain angle, the image of a pair of arms appeared.

Ramachandran asked the patient to look at the mirror image and move his phantom arm. What the patient moved was, of course, his existing arm, but the reflection sufficed to recalibrate his body map. After a week of training, the patient regained voluntary control over his phantom arm. Three weeks later, it disappeared from his body map, together with the pain. Ramachandran concluded:

> I realized that this was probably the first example in medical history of a successful 'amputation' of a phantom limb! [10]

The patient's brain was receiving intermodal incompatible signals when he saw his arm moving while his muscles signalled that there was no arm.

Ramachandran explained that the sensory conflict led the brain to discard the phantom arm altogether, rather than living with a conflicting body map [12].

This virtual obserpant extension through visual feedback from the mirror only "revived" the corresponding neural networks in the brain, as the actual limb no longer existed. However, this recalibration sufficed for the patient to regain a sense of agency.

Ramachandran's mirror therapy is also used in the rehabilitation of stroke victims, where re-mapping the brain allows patients to perceive paralyzed parts of the body and make them flexible again. Figure 3.4 shows an application of mirror therapy to treat phantom pain. It is widely applied today and portable rehab mirrors can be purchased in almost any mail-order business for use at home or on the move.

If a person with unilateral neglect is given a stick for reaching, this tool will open access to the hitherto neglected hemisphere, as the brain

Fig. 3.4. Ergotherapist Lynn Boulanger employs mirror therapy to treat the phantom pain of Marine Cpl. Anthony McDaniel in San Diego [13].

will interpret the stick as an extra limb. Through the haptic feedback provided by the stick as an elongation of his arm, the neglected part of the world will not only be within reach but also directly touchable. It is, of course, not touched directly but only via the stick. However, just as in our earlier example of the walking stick, the tool, together with the obserpant, merges into a systemic whole. Interestingly, the boundary towards the outside, which extends the obserpant, can be observed not only as behaviour but also as plastic changes on the neural level. This reversal is possible because our brain encodes our spatial environment by distinguishing extrapersonal space (far space) from peripersonal space (near space): Extrapersonal space is beyond reaching distance, and peripersonal space is in reach [14]. Those regions in the cortex which correlate with extrapersonal space are re-mapped in the brain as peripersonal space. The obserpant has (re-)captured parts of the environment which have become hard-wired as plastic neural changes. As a result, the previously reduced obserpant has regained a left hemisphere.

Berti and Frassinetti suggest that assigning objects or processes to the internal or external world is an adaptive, ongoing recalibration:

> … peripersonal and extrapersonal space coding is a dynamic process, not only related to the absolute coding of a distance, but also dependent on how the body extension is computed in the brain. [15]

These examples of extended spatial obserpants who have compensated spatial distance manifest themselves as local strong anticipative behaviour. The compensation involves a boundary shift which assigns part of what used to belong to the environment to the obserpant. Furthermore, the interface is transparent, while the obserpant retains a sense of agency.

In general, we may say that every time we compensate for a distance or delay between our previous extension and the newly acquired one, we perform a shortcut also referred to as local strong anticipation.

Adapting to environmental changes or to changes in our internal range of behaviour occurs via a two-way recalibration process. When we assign to ourselves phantom limbs and objects which previously belonged to our environment, as in the stick examples, we shift outward those boundaries which define us as a systemic whole. Conversely, when we assign to our environment an object which previously belonged to our self, we shift those boundaries inward. This is what happened to the neglected patients. The result is not anticipation but surprise.

Boundary shifts do not necessarily involve entire parts of the body. Reduced peripheral circulation, e.g. cold fingertips, may also be interpreted as a spatially reduced obserpant. Psychologist Hans IJzerman *et al.* discovered that social exclusion can result in a drop in finger temperature. The feeling of social exclusion can be remedied, however, by stimulating the cold fingertips through holding a cup of warm tea [16].

IJzerman *et al.* asked participants in an experiment to play *Cyberball*, a computer game in which people toss a ball to each other. During the game, the temperature of the index fingers which were not manoeuvering the mouse was measured. Subjects who did not receive the ball every two throws, i.e. who were socially excluded from the game, showed a decrease in their fingertips' temperature. For subjects who were included, i.e. who received the ball every two throws, no detectable change in temperature was measured [17].

In the second part of the experiment, "socially excluded" subjects, i.e. participants who had not received the ball often enough, were given a cup of warm tea to hold in their hands. Their negative feelings resulting from the social exclusion were alleviated. This was tested by asking the subjects to report whether they were feeling very or little stressed, tense, bad, etc. [18] IJzerman *et al.* extrapolate their findings:

> Physical warmth serves to repair the negative feelings of exclusion, and we suggest that people may use the perceptual system to interpret their social environment. We thus think that social exclusion leads to a decrease in skin temperature as an evolved simulator. [19]

The authors suggest that the link between physical warmth and affection is innate and may be a selection effect. Against the background of their theory of social thermoregulation, IJzerman *et al.* showed that mammals, including human beings, maintain their core body temperature through huddling, i.e. seeking social closeness to others [20]. The topic of social thermoregulation is picked up again in Chapter 4 as an example of predictive homeostasis.

Here, it shall serve as another example of how the interface between the obserpant and environment can be modified. The degree of social involvement can be measured in terms of peripheral temperature. Cold fingertips are less connected to the environment than warm ones, as less blood flow means reduced perception:

> A sense of disconnection may lead to vasoconstriction, a state in which the blood vessels are narrowed, maintaining body heat in the core, but not the periphery (like one's finger temperature). [21]

The degree of connectedness to the environment through peripheral body temperature is modifiable. The feeling of social disconnection with the resulting decrease in peripheral circulation, for instance, can be compensated through a physical increase in blood flow induced by environmental changes, such as a cup of warm tea. (Note that this compensation takes place only if the person has been conditioned to equate physical warmth with social warmth early in life.)

The compensation re-established the subjects' spatial extension and shifted the interfacial cut back towards the outside (their epidermis). Compared to the reduced state, the subjects had closed the gap between their physical extension (as measured in peripheral temperature) and that of their environment while retaining a sense of agency. Furthermore, as the core and periphery felt like one systemic whole again and the interfacial shift was transparent, they exhibited local strong anticipative behaviour [22].

3.4 Boundary Shifts: Temporally Extended Obserpants as Examples of Local Strong Anticipation

Strictly speaking, both spatial and temporal obserpant extensions are spatio-temporal extensions (see Chapter 12). To avoid unnecessary complications, I shall refer to temporal or spatial extension, depending on whether the concept of a delay or that of a distance dominates our perception of a situation.

An example of a temporal obserpant extension comes in the shape of delays which are inserted into a control loop and subsequently compensated. Philosophers Andy Clark and David Chalmers introduced the notion of an extended mind, which basically states that parts of the environment are coupled with the brain and assimilated [23]. They reassess an experiment by Carmena *et al.*, in which a macaque monkey's brain is connected to a computer via implanted electrodes [24]. The monkey watched and steered a cursor on a monitor by means of a joystick while the computer recorded the neural activity. When the experimenter disconnected the joystick, the animal continued to move the cursor through its neural activity. Then the control loop was extended by inserting a mechanical robot arm between the neural commands and visual feedback. The robotic arm caused friction which resulted in a 60–90 milliseconds delay. This at first confused the monkey, as it did not see the delay-inducing robot arm and received feedback to its neural commands only via the cursor. However, after a short period of recalibration, the macaque had compensated for the delay and continued to smoothly navigate the cursor across the monitor. Recalibration occurred also on the neural level, as the adaptation manifested itself in plastic reorganization of the cortex [25]. Clark concluded that the macaque had extended his mind.

The monkey had incorporated the robot arm, although it was placed outside its visual field, by compensating for the delay caused by friction. Transparent assimilated objects such as the robot arm which caused friction in the control loop give away their existence through the delay they cause within the control loop. But once the obserpant — in this case, the monkey — has compensated for the delay, full transparency and local strong anticipation govern the endo-perspective. Through incorporating the delay into its body schema, the macaque had become a temporally extended obserpant. From an outside vantage point, however, the inserted delay in the shape of the friction-inducing robotic arm remains visible [26]. In Section 10.3, we compare visible and transparent extensions as compensated and uncompensated perspectives in strongly anticipative systems.

A further example of a temporal obserpant extension is the compensation of the so-called "missing fundamental", a low musical pitch which is perceived by most human beings, even if it is physically not present, solely through its overtones (see Fig. 3.5):

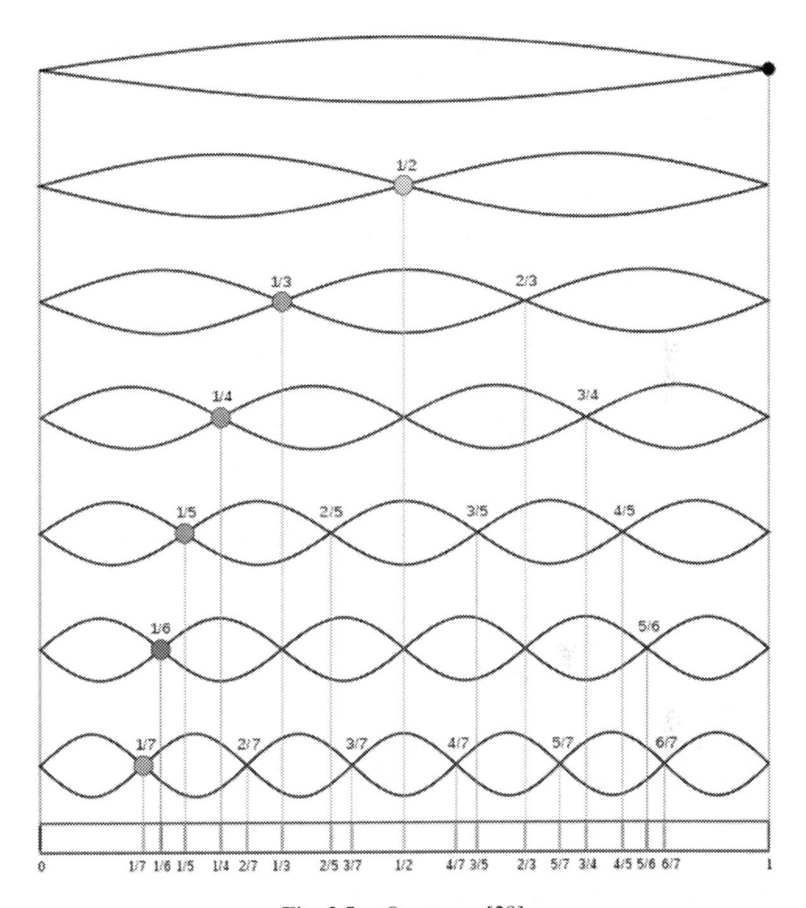

Fig. 3.5. Overtones [28].

> The fundamental frequency is the lowest in frequency of the harmonics (…) and it also has the greatest amplitude of all the harmonics. (…) the harmonics of a sound occur at progressive multiples of the fundamental frequency, e.g. 100, 100 × 2 = 200, 100 × 3 = 300, 100 × 4 = 400, and so on. If the fundamental frequency is removed, the pitch does not change. Even if the first three harmonics have been removed, one hears the tone as having the pitch of the original fundamental frequency, even when it is not physically present in the signal. [27]

Practical applications of the missing fundamentals are reduced overtones in loudspeakers and telephones. Small loudspeakers cannot produce low frequencies. So, if a note consists of a number of embedded overtones (200, 300, 400, 500, … Hz) which are integer multiples of the fundamental of 100 Hz and the fundamental is left out, the pitch which corresponds to the fundamental can still be heard, although it does not physically exist. Users of telephones which cannot produce frequencies of less than 300 Hz can still hear the lower frequencies, such as a male voice (about 150 Hz) although this frequency is not reproduced by the device [29].

The missing fundamental phenomenon is also used in diagnostics. Makiko Abe *et al.* showed that the impaired ability of Alzheimer's patients to perceive the missing fundamental is an indicator of the degeneration of brain sections involved in hearing [30].

Opinions differ as to how this phenomenon might be explained. The inferentialist position claims that the brain constructs the physically non-existent frequency through inference by applying an acquired rule. Autocorrelations have also been suggested to lie at the heart of this phenomenon, although there is little evidence of correlations on the neural level which may cause the phenomenon [31].

A recent convincing position by Judge [32] challenges the view that our brains represent our environment and construct the fundamental frequency. She suggests we do not need to assume that perceptual inference is at work:

> When a process appears to be rule-following, it could be the case that it is representing a rule and following it, or it could be merely co-varying with some environmental regularity, without representing that regularity. [33]

According to Judge, the missing fundamental may be explained as a result of the fact that our auditory system evolved together with an environmental regularity:

> When a given pattern of harmonics, A, consisting of components that are integer multiples of some frequency F, is encountered in the environment, this is almost always encountered as part of a complete complex tone with fundamental frequency F, of which A is a subset. (…) The system does not need to 'know' or to represent to itself the rule that this regularity obtains. It just needs to have evolved in a world where this

> regularity exists. We ought to consider that the functioning of a given perceptual system is determined, not just by the stimulus with which it is presented at a given time t, but with the types of stimulus with which it evolved to cope in the first instance. [34]

Judge's explanation is convincing. The following question, however, remains: Is the obserpant temporally extended if the longer amplitude of the missing fundamental — which embeds its overtones with shorter amplitudes — is perceived, albeit in a package with the remaining parts of the complex tone? An argument for the interpretation that the missing fundamental creates a temporal extension in the obserpant is that its wavelength is longer than those of its overtones and therefore spans a longer period of time.

The co-variation suggested by Judge manifests itself when we understand hearing as an active process, in which the ear emits frequencies which then form an interference pattern with incoming signals. Both emitted and received signals have evolved together from human and environmental coupling through mutual interaction. This position dovetails with Clark's claim that the act of hearing is itself an anticipatory process, as we expect to hear what we know from experience [35].

One can imagine the nature of the coordination between a co-evolving obserpant and environment either as a direct coupling of internal and external oscillations, an interference pattern or a coordination of long-term variations within the obserpant and within the environment. This differentiation is dealt with at length in Chapter 4 when we distinguish between strong anticipation as a result of local coupling as opposed to a global coordination of long-term fluctuations.

Another example of a boundary shift induced by auditory extensions is described by experimental musician Anthony Moore, who sees oto-acoustic emissions as an example of active perception. Such emissions are measurable sounds emitted by the ear, without which no sound could be perceived:

> … unless some action takes place on the part of the receiver, (…) incoming signals may remain unperceived. Due to the active, mechanical components that (…) make up the physiology of the ear, it is possible to

> construct a hypothesis which proposes that certain physical aspects of hearing take place outside the body, at the entrance to the ear. [36]

Physicist David Kemp discovered oto-acoustic emissions in 1978 [37]. He found that these emissions are of cochlear origin, resulting from the sensory hair cells responding to auditory stimuli. Nowadays, oto-acoustic emissions are recorded and analysed mainly for diagnostic purposes [38].

Sound is an interference pattern between oto-acounstic emissions and incoming signals. Moore denotes the interface — the extended boundary between the obserpant and environment — an "ethereal skin of interference", a membrane:

> ... active perception is the motive force that drives the efferent flow out through the portals of the body to meet the afferent flow of incoming waves from the physical world. What happens when these flows of information pass through each other is, in a series of cancellations and amplifications, what we come to know of as the world we move through. [39]

Oto-acoustic emissions are obserpant extensions which shift the interfacial cut outward and thus form a new systemic whole with incoming signals. As we are not aware of the extension (which includes waves outside the ear), it is transparent. Neither the ear nor the waves which belong to the environment need to be aware of the coupling. This auditory extension dovetails with the enactivists' "action-in-perception" credo, which assigns a proactive role to the listener and conveys a sense of agency. To change the metaphor, by stretching out one hand in the form of oto-acoustic emission, the resulting handshake is an example of local strong anticipative behaviour.

3.5 Boundary Shifts: Recalibrating Temporally Reduced Obserpants

Compensating a delay requires recalibration of the obserpant's perceptive apparatus. This may lead to an initial state of confusion before successful adaptation. Neuroscientist Chess Stetson *et al.* conducted experiments in which they inserted a delay between participants' keypresses and subsequent flashes on a screen. The participants adapted to the delay after an initial phase of temporal disorientation — they had compensated for the delay and thereby become temporally extended [40]. Next, the experimenters shortened the delay between keypresses and flashes. This caused the participants to experience an illusory reversal of cause and effect during recalibration: To them, the flashes appeared before the keypresses. The now temporally reduced participants also managed to adapt, this time through reintegrating the removed temporal extension.

Such recalibration as a reaction to delays preserves the obserpant's sense of agency, i.e. the feeling of causing, being in control of and executing one's actions. However, both a sense of agency and body ownership are necessary for bodily self-consciousness. Engineer and educationist Shu Imaizumi and psychologist Tomohisa Asai showed that although participants synchronized with a virtual hand after a delay had been inserted (thus retaining their sense of agency), no sense of body ownership was restored after visuomotor recalibration [41].

This means that, on the one hand, the short delay between motor action and visual feedback which is compensated through temporal recalibration appears not to compromise a sense of agency. On the other hand, a sense of body ownership is not elicited after delay compensation. The authors suggest that the sense of agency and body ownership are dissociated after visuomotor temporal recalibration.

Compensating inserted and removed delays and distances may thus have profound consequences for us. Recalibration is an adaptation, not a reversal of performed compensation. Chapter 11 discusses such adaptations against the background of a range of disorders and diseases which result from successful or unsuccessful compensation.

To conclude, compensated spatial distance and compensated delays, both of which result from obserpant extensions, are examples of local strong anticipation if the following conditions are met: The extension of the new systemic whole transcends the original interfacial cut between the obserpant and the rest of the world, the boundary shift itself remains transparent to the obserpant, and a sense of agency is preserved.

The following chapter deals with strong anticipation as delay compensation, with examples of anticipatory synchronization and the coordination of long-term correlations within the living and inanimate world.

References

[1] M. Botvinick and J. Cohen, Rubber hands 'feel' touch that eyes see. *Nature*, 391, 756, 1998.

[2] A. Clark, *Supersizing the Mind. Embodiment, Action and Cognitive Extension*, Oxford University Press, New York, 2008, pp. 30–37.

[3] S. Ivaldi, Intelligent Human-Robot Collaboration with Prediction and Anticipation. Ercim.news, 2018. https://ercim-news.ercim.eu/en114/special/intelligent-human-robot-collaboration-with-prediction-and-anticipation.

[4] S. Ivaldi, Anticipatory models of human movements and dynamics: The roadmap of the AnDy project. 2017, p. 7. https://hal.science/hal-01539731/document.

[5] Reproduced with permission from Serena Ivaldi and Francesco Nori (Edited, original in colour). (Serena Ivaldi, Anticipatory models of human movements and dynamics: the roadmap of the AnDy project. 2017, p. 7. https://hal.science/hal-01539731/document).

[6] O. Sacks, *The Man Who Mistook his Wife for his Hat*, Picador, 1986, pp. 53–54.

[7] V. S. Ramachandran and S. Blakeslee, *Phantoms of the Brain*, Fourth Estate, London, 1999.

[8] Photograph by B. Ring, Dr. Vilayanur S. Ramachandran and psychology student Matthew Marradi holding the original Mirror Box.jpg. Taken on 2. Mai 2012, licensed under CC BY-SA 3.0. https://commons.wikimedia.org/wiki/File:Dr._Vilayanur_S._Ramachandran_and_psychology_student_Matthew_Marradi_holding_the_original_Mirror_Box.jpg (original in colour).

[9] V. S. Ramachandran and S. Blakeslee, *Phantoms of the Brain*, Fourth Estate, London, 1999, p. 46.

[10] *ibid*, p. 49.

[11] By Phidauex, own work, public domain, https://commons.wikimedia.org/w/index.php?curid=1071975 (edited, original in colour).

[12] V. S. Ramachandran and S. Blakeslee, *Phantoms of the Brain*, Fourth Estate, London, 1999, pp. 49–50.

[13] U.S. Navy Photography of Mass Communication Specialist Seaman Joseph A. Boomhower. 2011 (This file is a work of a sailor or employee of the U.S. Navy, taken or made as part of that person's official duties. As a work of the U.S. Federal government, it is in the public domain.) https://commons.wikimedia.org/wiki/File:US_Navy_110613-N-YM336079_Lynn_Boulanger,_an_occupational_therapy_assistant_and_certified_hand_therapist,_uses_mirror_therapy_to_help_address_phan.jpg.

[14] A. Berti and F. Frassinetti, When far becomes near: Remapping of space by tool use. *Journal of Cognitive Neuroscience*, 2(3), 418–419, 2000.

[15] *ibid*, pp. 418–419.

[16] H. R. IJzerman, M. Gallucci, W. T. J. L. Pouw, S. C. Weiβgerber, N. J. Van Doesum, K. D. William, Cold-blooded loneliness: Social exclusion leads to lower skin temperatures. *Acta Psychologica*, 140, 283–288, 2012.

[17] *ibid*, p. 286.

[18] *ibid*, p. 286f.

[19] *ibid*, p. 286.

[20] H. R. IJzerman, J. A. Coan, F. M. A. Wagemans, M. A. Missler, I. van Beest, S. Lindenberg and M. Tops, A theory of social thermoregulation in human primates. *Hypothesis and Theory, Frontiers in Psychology*, 6, 12, 21 April 2015.

[21] H. R. IJzerman, M. Gallucci, W. T. J. L. Pouw, S. C. Weiβgerber, N. J. Van Doesum, K. D. Williams, Cold-blooded loneliness: Social exclusion leads to lower skin temperatures. *Acta Psychologica*, 140, 284, 2012.

[22] Note that this is my interpretation, not necessarily that of the authors.

[23] A. Clark, *Supersizing the Mind. Embodiment, Action and Cognitive Extension*. Oxford University Press, New York, 2008, pp. 30–37.

[24] J. Carmena, M. Lebedev, R. Crist, J. O'Doherty, D. Santucci, D. Dimitrov, P. Patil, C. Henriquez, and M. Nicolelis, Learning to control a brain-machine interface for reaching and grasping by primates. *Public Library of Sciences: Biology*, 1(2), 193–208, 2003.

[25] *ibid*, pp. 193–208.

[26] S. Vrobel, The extended mind: Coupling environment and brain. In: In D. M. Dubois (ed.) *Computing Anticipatory Systems: CASYS '09: Proceedings of the Ninth International Conference on Computing Anticipatory Systems*, AIP Conference Proceedings, Vol. 1303, November 2010, pp. 277–282.

[27] https://homepage.ntu.edu.tw/~karchung/Phonetics%20II%20page%20thirteen.htm. Accessed 4.3.2023.

[28] Moodswingerscale.jpg. https://commons.wikimedia.org/wiki/File: Moodswingerscale.svg, public domain.

[29] https://en.wikipedia.org/wiki/Missing_fundamental. Accessed 3.3.2023.

[30] M. Abe, K.-I. Tabei, M. Satoh, M. Fukuda, H. Daikuhara, M. Shiga, H. Kida and H. Tomimoto, Impairment of the missing fundamental phenomenon in individuals with Alzheimer's disease: A neuropsychological and voxel-based morphometric study. *Dementia and Geriatric Cognitive Disorders*, 8(1), 23–32, 2018.

[31] https://en.wikipedia.org/wiki/Missing_fundamental.

[32] J. A. Judge, Does the 'missing fundamental' require an inferentialist explanation? *Topoi*, 36(2), 319–329, 2017.

[33] *ibid*, p. 325.

[34] *ibid.*

[35] A. Clark, *Surfing Uncertainty. Prediction, Action and the Embodied Mind*, Oxford University Press, 2016/2019.

[36] A. Moore, Transactional fluctuations 1 — Towards an encyclopaedia of sound. In: S. Zielinski and E. Fürlus (eds.) *Variantology 3*. Verlag der Buchhandlung Walther König, Köln, 2008, p. 299.

[37] D. T. Kemp, Stimulated acoustic emissions from within the human auditory system. *The Journal of the Acoustical Society of America*, 64(5), 1386–1391, 1978.

[38] D. T. Kemp, Oto-acoustic emissions, their origin in cochlear function, and use. *British Medical Bulletin*, 63(1), 223, 2002.

[39] A. Moore, Transactional fluctuations 1 — Towards an encyclopaedia of sound. In: S. Zielinski and E. Fürlus (eds.) *Variantology 3*. Verlag der Buchhandlung Walther König, Köln, 2008, p. 301; and personal communication, 2009.

[40] C. Stetson, X. Cui, P. R. Montague and D. M. Eagleman, Illusory reversal of action and effect. *Journal of Vision*, 5, 769a, 2006.

[41] S. Imaizumi and T. Asai, Dissociation of agency and body ownership following visuomotor temporal recalibration. *Frontiers in Integrative Neuroscience*, 9, 35, 2015.

Chapter 4

Strong Anticipation as Delay Compensation

Local strong anticipation for extended obserpants manifests itself in a boundary shift which results in the formation of a new systemic whole, transparent interfaces and a sense of agency. By contrast, local strong anticipation in anticipatory synchronization unfolds through enslaved systems which synchronize with master systems. Both the conditions for strong anticipation in extended obserpants and in anticipatory synchronization imply a compensatory act by one component of the newly formed systemic whole.

Local strong anticipation in a system embedded in at least two coupled environmental rhythms emerges through indirect contact. If that system is entrained with the shorter nested rhythm, it is also indirectly embedded in the longer one [1]. This indirect contact with a temporally more extended rhythm can provide the anticipating system with information which was previously inaccessible for it, as it belonged to that system's future.

By locking into two or more spatio-temporally embedding environmental systems, the anticipating system is thereby embedded in a temporally more extended process. This generates a spatio-temporal gap which is bridged, i.e. compensated, by the anticipatory system [2]. Local strong anticipation for extended obserpants, for anticipatory synchronization and for indirect coupling can be understood as a general principle of delay and distance compensation.

Global strong anticipation results from the coordination of long-term correlations between fluctuations on multiple time scales on both sides of the interfacial cut. In this relation, delay and distance compensation are not as easily detectable as in direct coupling between internal and external rhythms. Rather, it is the fluctuations, i.e. the way an aspect of the world changes, which are coordinated. A candidate for such dynamics, $1/f$ (or pink) noise, is introduced in Chapter 5.

This chapter deals with strong anticipation in the various forms of delay and distance compensation. Before we touch on the compensation of delays in physical and biological systems, though, it is necessary to remind ourselves of neurocognitive delays we compensate unwittingly and which always precede the compensation of any environmental delays as "add-ons".

4.1 Compensating Cognitive and Neurocognitive Delays: Anticipation, Postdiction and Peri-diction

Some delays occur within the human brain and body, while others result from environmental constraints. When it comes to human perceptions and conscious awareness, it is important not to confuse two types of delay compensation: anticipatory and postdictive. Anticipatory delay compensation occurs, for instance, in the visual and motor system when an outfielder catches a ball in flight. Postdictive compensation relates to cognitive and neurocognitive delays.

The question at what stage between perception and cognition delay compensation takes place, however, is still a matter of debate. Psychologist Romi Nijhawan suggests that visual prediction occurs in sensory pathways, so when we perceive motion, this perception consists of already compensated delays [3]. He claims that, during the act of catching a ball in flight,

> … the visual system modifies the perceived position of the ball so that it matches the position of the moving hand. [4]

Eagleman suggests a complementary method to Nijhawan's position that anticipation, i.e. delay compensation, happens at the level of the perceptual system. He points out that, in addition to anticipative mechanisms, the brain implements postdictive strategies [5].

Postdictive delay compensation occurs when the brain makes sense of events in the outside world. When we see, hear, smell or touch, sensory information does not instantly reach our brains nor is it instantly processed. Neuroscientist David Eagleman suggests that our brains coordinate all sensory information in order to put together the best story about the outside world it can construct:

> Your brain, after all, is encased in darkness and silence in the vault of the skull. Its only contact with the outside world is via the electrical signals exiting and entering along the super-highways of nerve bundles. Because different types of sensory information (hearing, seeing, touch, and so on) are processed at different speeds by different neural architectures, your brain faces an enormous challenge: what is the best story that can be constructed about the outside world? [6]

In order to create a convincing story, the brain waits, within a certain window, for the slowest information to arrive and then creates a consistent perception of the world. For visual perceptions, this window spans about a tenth of a second. The fact that our visual system makes up for delays is well known:

> In the early days of television broadcasting, engineers worried about the problem of keeping audio and video signals synchronized. Then they accidentally discovered that they had around a hundred milliseconds of slop: As long as the signals arrived within this window, viewers' brains would automatically resynchronize the signals; outside that tenth-of-a-second window, it suddenly looked like a badly dubbed movie. [7]

Eagleman concludes that this window of delay is evidence that conscious awareness is postdictive rather than anticipatory. In a way, we live in the past when our conscious awareness is concerned. He exemplifies this unintuitive fact with a thought experiment: Touch your toe and your nose at the same time and you will perceive this action as simultaneous, although the signal from your toe takes longer to reach your brain. It seems your brain waits for a while to see if there are any more signals to arrive and then synchronizes them:

> It may be that a unified polysensory perception of the world has to wait for the slowest overall information. Given conduction times along limbs, this leads to the bizarre but testable suggestion that tall people may live further in the past than short people. [8]

Eagleman stresses that we have to keep in mind that postdiction belongs to conscious phenomena only. Motor reflexes cannot afford to wait if, for instance, a fly hits your eye:

> It must be emphasized that everything I've been discussing is in regard to conscious awareness. It seems clear from preconscious reactions that the motor system does not wait for all the information to arrive before making its decisions but instead acts as quickly as possible, before the participation of awareness, by way of fast subcortical routes. [9]

Sometimes, both anticipative and postdictive delay compensation are at work. An example is phenomena such as spatial warping which compensates for neural delays.

In an experiment known as the Hering Illusion, neuroscientists Don Vaughn and David Eagleman set out to test the "perceive the present" (PTP) hypothesis, which was suggested by theoretical cognitive scientist Mark Changizi [10] (see Figs. 4.1 and 4.2). It claims that some geometric illusions are caused by delays which the visual system has to compensate for. Rather than perceiving a slightly delayed conscious image of the world (approximately 100 ms in the past), our visual system anticipates what the world will look like in the near future [11]. In the case of the Hering Illusion,

> the PTP hypothesis proposes that the background of radial lines simulates optic flow, causing the visual system to assume forward egomotion and to extrapolate the appearance of the parallel bars to the next moment. Because objects closest to the horizontal plane move fastest

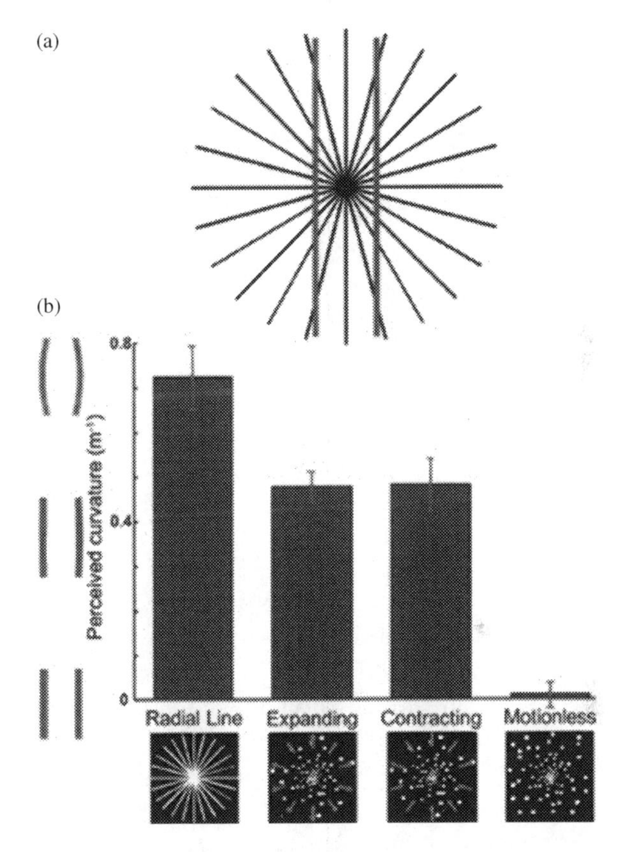

Fig. 4.1. Participants viewed two bars flashed above a background of radially expanding or contracting dots. In randomly interleaved trials, radial lines or a control background of motionless dots was used. The bars were flashed for 80 ms with an interstimulus interval of 1 s [13].

> during forward motion, this generates the illusory percept that the two parallel bars bend outward. Imagine driving on a suspension bridge toward two of its pillars: from a distance the pillars appear as parallel lines. As you approach, the pillars move farther apart at eye level, but their distant tops still appear close together. [12]

Vaughn and Eagleman questioned whether the mechanisms were predictive, suspecting that they might also be postdictive. To test this, participants viewed

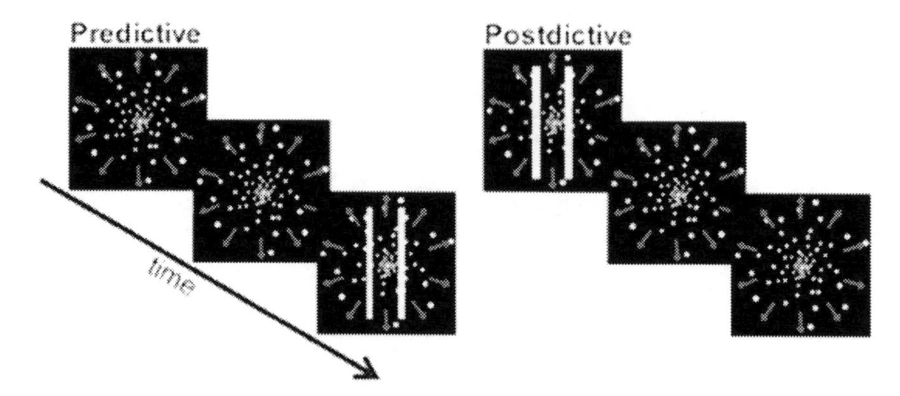

Fig. 4.2. The magnitude of the illusion is identical whether background motion precedes the presentation of the bars (prediction) or follows it (postdiction). Results reveal a window of motion integration between 80 and 160 ms. In both conditions, the bars flashed for 80 ms [15].

> … a 1 s expanding optic flow pattern offset-aligned with 80 ms bars (predictive case) or onset-aligned (postdictive case). If optic flow induces spatial warping by extrapolation, any optic flow after the presence of the bars should have no effect on the illusion magnitude. We found, in contrast, that information collected in a ~80 ms window on either side of the bars contributes equally to the spatial warping (…). In other words, the effect is not merely postdictive or predictive, but symmetrically peri-dictive: there is a symmetrical temporal window of motion integration around the flashing of the bars. [14]

As the visual distortion is the same whether the optic background flow preceded or followed the flashing of bars, the researchers concluded that their findings show that spatial warping is "peri-dictive", i.e. both anticipatory and postdictive.

To compensate for delays, both anticipative and postdictive mechanisms are at work in order to construct a consistent world. Anticipation helps us to navigate the world smoothly as it allows us to react more quickly to environmental and internal changes. But is postdiction relevant to anything besides our subjective consciousness? The answer is yes, in particular, if the causal order appears to be compromised by the removal of an existing delay. This was shown by Stetson *et al.*'s experiment in which an existing delay between a keypress and a flash appearing on a screen was removed (see Chapter 3). Individuals who participated in the experiment reported that after the removal of the delay, the causal order of events was reversed: They saw the flashes before they pressed the key. This reversal occurred because the participants anticipated a longer delay. Eagleman stresses the importance of this reversal:

> Note that the recalibration of subjective timing is not a party trick of the brain; it is critical to solving the problem of causality. At bottom, causality requires a temporal order judgment: did my motor act come before or after that sensory signal? The only way this problem can be accurately solved in a multisensory brain is by keeping the expected time of signals well calibrated, so that "before" and "after" can be accurately determined even in the face of different sensory pathways of different speeds. [16]

It is vital for us to be able to assign perceptions, without doubt, either to our environment or to ourselves. However, during recalibration, it is often not possible to confidently assign a perception to either side of the obserpant-world interface. Stetson *et al.* provide an outside-of-the-lab example of the importance of correct assignments:

> Imagine you are walking through a forest and hear a twig crack. Did it happen when your foot fell or just before? If it happened just before, the sound may alert you to a nearby predator. If the sound happened coincident with your step, then it was a normal occurrence consistent with the sensory feedback expected during walking. [17]

4.2 Transparency as Delay Compensation

Another type of delay compensation which has to be considered an "add-on" is transparency. An internal model entails the idea of representations of the external world and of an agent who interacts with his or her environment. If such a model is transparent to the obserpant, it has the advantage of saving time and energy, as an interpretation or compilation of a representation may be skipped and an all-too-close encounter with a bear might be evaded. A truly viable wired-in model is invisible: In a phenomenological approach, philosopher Thomas Metzinger claims that our internal models of ourselves and our environment are transparent to us. We are not aware of the creation of a representation of the external world and ourselves within our brains and tend to insist that the contents of our phenomenal self-model are what we perceive. There is no lasting entity like a "self", but just a "phenomenal self", which makes this phenomenal self, including its environment, seem very real to us [18]. Metzinger concludes that this transparency of our perceptual apparatus, this inability to identify percepts as representations, is probably a selection effect.

We can react in time in the face of lurking danger because our representations of the outside world are transparent to us. We do not recognize them as representations, which means we perceive the world as naïve realists, a perspective which, according to Metzinger, is extremely useful for dealing with the world, for instance, in the face of an approaching predator:

> ... naïve realism has been a functionally adequate assumption for us, as we only needed to represent the fact 'there's a wolf there' (...) not 'there's an active wolf representation in my brain now'. [19]

This phenomenal transparency, which adds an additional degree of anticipation to our perception and cognition of the world, is a matter of degree. Full transparency exists if the representational character is not accessible to subjective experience. The other extreme, opacity, exists when we are aware of the fact that something is a representation. Metzinger describes transparency as a "special form of darkness", whereas opacity "appears when darkness is made explicit" [20].

Unlike local strong anticipation in extended obserpants, phenomenal transparency does not involve an outward interfacial shift to include a part of the environment. However, transparency is a property which makes the experience of the world and oneself more immediate than an opaque experience in which we are aware of our internal representational model. By skipping unnecessary or cumbersome delays, we perform a compensatory act which qualifies as a manifestation of anticipation.

Although they help us to navigate the world smoothly, compensatory acts, even those in the form of our phenomenal transparency, are a constraint on our epistemological endeavours. Only if "darkness is made explicit" can we reveal compensated temporal and spatial extensions outside our current awareness.

4.3 Simultaneity Horizons as Boundaries of Delay Compensation

After this brief excursion into internal intermodal delays and phenomenal transparency, this section focuses on our temporal alignment with the outside world. Events become perceivable or registrable if they propagate signals in the form of air pressure waves, light waves or disturbances of any other kind. Such signals, however, are not immediately perceived by a human being or registered by a device without delay, as signal propagation and processing take time. As Eagleman has pointed out, the phenomenon of differing propagation speeds and processing times has been known for a long time and taken advantage of in all walks of life:

> A gun is used to start sprinters, instead of a flash, because you can react faster to a bang than to a flash. This behavioral fact has been known since the 1880s and in recent decades has been corroborated by physiology: the cells in your auditory cortex can change their firing rate more quickly in response to a bang than your visual cortex cells can in response to a flash. [21]

In addition to the difference in processing time, propagation speed also differs. Thunder and lightning are generated by the same external event, but as the speed of light exceeds that of sound, the signals arrive in succession if we are located far from the storm and do not compensate for the delayed signal.

Psychologist and neuroscientist Ernst Pöppel defined a "simultaneity horizon" at 10–12 meters distance from an obserpant, at which auditory and visual signals are perceived as occurring simultaneously because all delays are compensated. At this horizon, the different delays in the propagation speed of sound and light are compensated through the difference in our processing time of visual and acoustic stimuli. Within and beyond Pöppel's simultaneity horizon, visual and acoustic stimuli are no longer perceived as simultaneous but as two successive events. Within the horizon, acoustic stimuli arrive before visual ones, and beyond it, visual stimuli reach us before acoustic ones [22].

Actions and perceptions inside and outside us take time to reach our consciousness. Pöppel's simultaneity horizon marks the point at which the auditive and visual world merge into a coherent systemic whole with the obserpant. As new systemic wholes emerge when internal and external delays are compensated, we may define — for our purposes — a systemic whole as a system situated at the simultaneity horizon.

The following section deals with concrete examples of compensated external delays as observed in coupled physical, biological and social systems. While feedback delays within a system or between a system and its environment were traditionally regarded as error-inducing annoyances, they are now widely regarded as the key to anticipatory synchronization and regulation which stabilizes the balance between a system and its environment (or between two systems) [23]. In contrast to the concept of local strong anticipation in extended obserpants, anticipatory synchronization, as defined by Stepp and Turvey's five necessary conditions (see Section 2.6.1), rejects the idea of internal representations.

4.4 Local Strong Anticipation in Physical Systems as Delay Compensation

Physicist and mathematician Christiaan Huygens synchronized two pendulum clocks hanging from a board that was supported by two chairs. The spontaneous coupling of the clock's phases was brought about and determined by the physical consistency of the board and the differences between the individual clock's phases [24].

Today, synchronization through phase-locking or entrainment is often used to stabilize physical systems. Apart from spontaneous entrainment, phase-locking can be steered through anticipatory synchronization. It occurs when a "slave system", which is embedded in and coupled to a "master system", synchronizes with the master system's behaviour after the insertion of a small temporal feedback delay.

A classic example is the synchronization of light particles in a laser beam. Lasers were first portrayed as a self-organizing process in the 1970s by theoretical physicist Hermann Haken in his seminal book "Synergetics." Here, Haken introduces the notions of "slave", "enslavement" and "master" (which he denotes as "order parameter") [25]. Synergetics studies the interaction of nonlinear subsystems which self-organize into larger systemic wholes. It describes both the microscopic and macroscopic behaviour of a system and the way the two levels interact. Lasers can be synchronized, for instance, by means of injection locking. In a so-called "phase synchronization", two or more continuous-wave lasers are coupled when the output of the master laser (or seed laser) is injected into the slave laser(s). The master is a single-frequency laser which forces the slave to emit a phase-synchronized optical frequency.

Delay-induced anticipatory synchronization in coupled lasers can turn chaotic behaviour into stable dynamics. Here, the slave (also called receiver or response system) anticipates the state of the master (alternative terms: transmitter or drive system), although the delay was introduced into the master. Other examples of delay-induced anticipatory synchronization abound in the literature [26]. Building on Dubois' extensive research on anticipatory synchronization in chaotic systems, this phenomenon has been observed in many disciplines.

Physicist Henning U. Voss showed that time-delayed feedback can induce anticipating synchronization:

> ... dissipative chaotic systems with time-delayed feedback can force identical systems onto a synchronization manifold that involves the future state of the drive system, and (...) those systems can synchronize in the usual sense even if the coupling between both lies in the past. These results are counterintuitive, since in both cases the future evolution of the drive system is anticipated. [27]

He describes anticipating synchronization as the process

> where the response system anticipates the drive system, just by synchronizing to a state that lies in the future of the driving system. [28]

According to Turvey, anticipating synchronization through delay coupling between a changing environmental feature (x) and a perception-action system (y) can be expected if their intrinsic dynamics (f and g) are similar [29]. If applied to organisms with oscillating dynamics which are enslaved by circadian rhythms, such as the behaviour of slime mould (see the following section), frequency locking can be expected if their periods are similar:

$$\dot{x} = f(x)$$
$$\dot{y} = g(y) + k\,(x - y_{\tau})$$

where k is the coupling strength and y_{τ} is $y(t\text{-}\tau)$ that is a past state of y delayed by τ. [30]

Anticipatory synchronization is an example of local strong anticipation. In fact, according to Stepp and Turvey, the two terms are identical [31]. Turvey points out that strong anticipation and direct perception both reject predicting models, are both based on lawful relations and are thus almost interchangeable terms:

> ... at a minimum, Strong Anticipation is a generalization of Direct Perception. At a maximum (...) the two conceptions are one and the same. [32]

4.5 Local Strong Anticipation in Biological Systems as Delay Compensation

Anticipative behaviour in biological systems has been described for a long time. However, as biologist Carrie Deans pointed out, these descriptions often had no association with anticipation [33]. Examples are anticipative salivation in dogs [34] and entrainment to circadian rhythms [35], which today would be denoted anticipatory regulation or anticipatory synchronization. The idea is simple:

> When environmental factors change in predictable ways, the detection of a change in one variable can reliably predict a future change in another. [36]

The underlying principle at work in this type of anticipatory regulation is the fact that an organism, or components thereof, is coupled with an environmental factor which is itself correlated to yet another environmental factor. If the first factor (to which the organism is coupled) temporally precedes the first one, the organism can prepare its response well ahead of the impending environmental impact. This section describes anticipatory regulation in simple and complex biological systems.

Deans pointed out that biogeochemical cycles in our environment often produce reliable spatio-temporal patterns and environmental factors are often strongly correlated with each other [37]. This means that apart from engaging in reactive responses, systems can anticipate the temporal development of environmental factors if asynchronous signals are related to these factors. In Fig. 4.3, Deans illustrates the difference between the way a reactive and an anticipative system relate to environmental factors:

> Figure [4.3] provides a graphical representation of how these different signals can affect organismal responses. The time course of a theoretical environmental factor is shown in panel (A), and panel (B) shows the time course of Signal 1, a signal with the same temporal pattern as the environmental factor, and Signal 2, which is a signal that is correlated with, but precedes, the environmental factor. (…) C shows that because the reactive response is regulated by Signal 1, there is a lag in the response time. This delay is associated with the amount of time it takes to produce the response, such as time related to gene transcription and

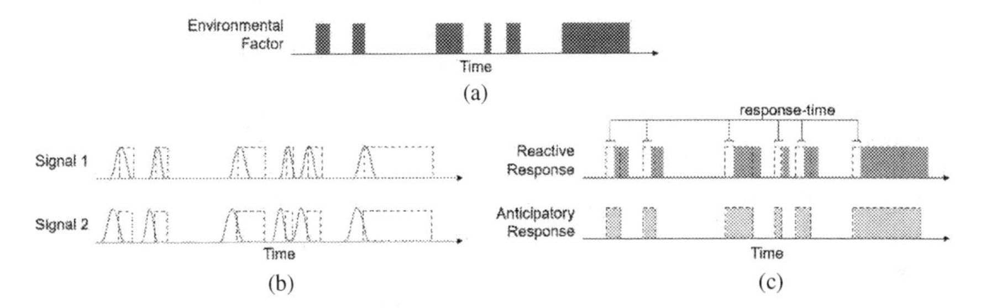

Fig. 4.3. Example showing the time course of a hypothetical environmental factor (A), the time course of two signals that are correlated with this variable (B) and a demonstration of how the temporal relationship between the environmental variable, signal and response can be used to distinguish between reactive and anticipatory processes (C) [39].

> protein production/transport. The anticipatory response, on the other hand, has no lag in response time because it is regulated by Signal 2, which precedes the change in environment. [38]

4.5.1 *Synchronization through coupling*

Biological systems, from amoebae to human beings, display anticipatory behaviour in the form of synchronization through coupling with their respective environments. Coupling between the organism and environment allows the organism to coordinate with the future.

Biologist Aurèle Boussard *et al.* describe how oscillations in slime mould (*physarum policephalum*) result in adaptive behaviour and anticipation through the synchronization of oscillations of various lengths. Both mechanical and chemical oscillations interconnect and synchronize so as to display coordinated behaviour [40]. One such oscillator is the relaxation–contraction cycle, which transports cytoplasm within the network, with rather short periods of approximately 90s. These are embedded in oscillations with longer periods, which govern mitosis (a few hours) [41]. Boussard *et al.* describe how synchronization arises from two or more interacting oscillators:

> The position of an oscillator in its cycle can be described in terms of its phase, which advances at the pace set by the oscillator's natural frequency. Interacting oscillators influence each other's phase progression (...). A paradigm for synchronization is a population of globally coupled oscillators (i.e. each oscillator in the population is affected by the dynamics of the rest), in which the natural frequencies are initially heterogeneously distributed. As time progresses, the oscillators minimize their pairwise phase differences, resulting in the pattern known as full synchronization, where all oscillators operate at the same instantaneous frequency and phase. [42]

Slime mould responds to environmental stimuli by adjusting its oscillation, first locally and then globally [43]. This corresponds to a general mechanism, which manifests

> ... environmentally driven behavioural change (...) as integration of environmental information through change in the oscillatory pattern. [44]

Biologist Tetsu Saigusa *et al.* showed that the true slime mould anticipates periodicity in environmental changes and developed a dynamical systems model which simulates the anticipative behaviour [45]. The researchers removed the tip of a large plasmodium of slime mould, placed it under constant temperature and humidity on a narrow lane and started an experiment:

> The organism migrated along the lane, and the position of its tip was measured every 10 min. After migration had been permitted for a few hours, the ambient conditions were changed to cooler (…) and drier (…), referred to as 'dry stimulation' conditions for 10 min (…). This dry stimulation was repeated 3 times (…). [46]

They found that during dry periods, locomotion slowed down. When the original humidity level was restored, locomotion slowed down at precisely those moments at which the next dry stimulation was expected. Saigusa *et al.* conclude that this behaviour signifies the anticipation of upcoming environmental change. The periodic change between warm/dry and cooler/dryer cycles provided the enslaving master system to which the dynamics of the slime mould synchronized.

4.5.2 *Anticipatory regulation by eavesdropping*

An example of anticipatory regulation in plants is their behaviour towards environmental cues known as eavesdropping. If plants are stressed through damage, they send out warning cues about impending threats such as herbivory or drought in the form of volatile or underground cues. If nearby unstressed plants perceive these stress responses, they relay them, thus eliciting stress responses in other unstressed plants. Thus, plants can respond to relayed cues, which are emitted before actual damage occurred by producing airborne or underground chemicals which protect them against damage and also warn their neighbours. The alternative, reacting to stress cues produced by damaged plants, would leave less time to prepare for the impending danger. Thus, eavesdropping allows plants to anticipate forthcoming stress [47].

Eavesdropping between different plant species is described by entomologist Richard Karban *et al.* Over five seasons, they conducted experiments on the behaviour of tobacco plants in the vicinity of sagebrush, which was clipped so as to induce stress responses. The clipped sagebrush then released polyphenol oxidase (PPO), which to many herbivores is an anti-nutritive protein [48]. As a result, tobacco plants near clipped sagebrush showed more PPO activity and suffered less leaf loss through herbivores. Real herbivory had the same effect of less leaf damage in neighbouring tobacco plants.

The researchers suggested that eavesdropping between the two different species had taken place, which led to anticipative behaviour in the tobacco plants. They based this claim on the observation that both tobacco plants and sagebrush share a common herbivore: the grasshopper. However, as sagebrush grows faster, the insects feed on it early in the season. When tobacco plants are available later in the season, the grasshoppers feed less on sagebrush. Karban *et al.* conclude the following:

> Presumably tobacco responds early in the season and early in its development to airborne cues released by sagebrush. When neighbouring sagebrush experiences damage early in the season, tobacco in the vicinity may be more likely to be damaged later. The difference in timing of damage may make sagebrush a particular good source of information about risk of herbivory for neighboring tobacco. [49]

In this case, the anticipative behaviour of the tobacco plant could only be described if all three systems — the tobacco plant, the atmosphere and the sagebrush — formed one systemic whole. It resulted from the coupling of two plant species within a mediating environment, which allowed for airborne transmission of chemical substances. A step was "skipped" in the tobacco plants' reaction, as they did not have to wait for the actual herbivory to produce chemical defence but were able to generate it well in advance of the imminent danger. A compensated delay manifests itself in the "skipping", which allows a reaction well in advance of the actual threat of herbivory.

Deans points out that plant eavesdropping often leads to priming in plants. Priming simply means exposing an organism to an environmental stimulus in order to strengthen the organism's response in later encounters with the same stimulus. Like vaccinations produce an improved immune response, priming produces an improved direct response.

Although response through priming is direct, i.e. not asynchronously coupled to environmental stimuli, Dean classifies it as anticipatory behaviour:

> Given that priming is implemented as a response to future scenarios and it increases biological efficiency by reducing response time, it qualifies as an anticipatory process. [50]

Priming may well be classified as anticipative behaviour, as the plant is better prepared for future encounters with, e.g. a pathogen because it carries a model of the pathogen inside itself. However, it is not a case of local strong anticipation. It neither fulfils Stepp and Turvey's conditions for anticipatory synchronization nor does priming meet my four conditions for strong anticipation in extended obserpants.

4.5.3 *Epigenetics*

A transgenerational manifestation of strong anticipation needs to be considered when we look at extensive temporal periods which span not only the time from the first to the second encounter of a plant with the same pathogen. Such temporal extensions would encompass seasons, years and possibly even longer time spans, if we consider priming across generations.

Defensive responses plants produced under herbivore attack can also be found in their offspring. It is induced through epigenetic responses such as DNA methylation [51]. Epigenetic responses are processes which change gene expression without changing the DNA sequence. Unlike genetic modification, epigenetic changes can evolve within a lifetime and can be passed on to an organism's offspring.

Deans points out that epigenetic processes differ from genetic ones in terms of their stimulus–response relation. Resembling stress priming and circadian rhythms, epigenetic processes can serve as anticipatory biological mechanisms:

> While genetically mediated processes are often directly triggered by environmental stimuli, epigenetic processes appear to be primarily mediated by internal signals. The information held and transmitted in epigenetic marks do not contain any actual information about environmental patterns, but rather information about past transcription responses to environmental conditions. In many ways, this is similar to circadian rhythms, which are initiated by an internal clock that is periodically entrained by environmental stimuli. Here, rather than constantly coordinating biological processes with external signals, circadian-regulated processes respond more efficiently to internal cues that model predictable environmental changes. [52]

An example of transgenerational priming is evolutionary ecologist Mar Sobral *et al.*'s findings that caterpillar herbivory changed the plant epigenome as well as chemical and physical defences both within and across generations in wild radish. The responses differed, depending on developmental stages and whether the defence was of a chemical or physical nature:

> Across generations, herbivory experienced by mother plants caused strong direct induction of physical defenses in their progeny, with effects lasting from seedling to reproductive stages. For chemical defenses, however, this transgenerational induction was evident only in adults. [53]

Epigenetic modifications are not the only way organisms can acquire anticipatory models. Philosopher and biologist Matthew Sims considers four ways in which anticipatory models may be acquired, both over phylogenetic and ontogenetic time scales: mutation-based, epigenetic inheritance-based, associative learning-based and Baldwin effect-based model acquisition [54].

Mutation-based model acquisition occurs if

> some subpopulation of organisms comes to capture the correlational structure of the environment after repeatedly encountering that structure over multiple generations via (…) permanent modification in the DNA sequence of a gene (…). Natural selection increases the random

> mutation in the population until most organisms have acquired the anticipatory model. [55]

However, as Sims points out, not all heritable variation is a result of genetic change. Epigenetic inheritance-based model acquisition results not from genetic mutations but from variations resulting from environmental or internal regulatory factors during the organism's lifetime. As a result, offspring may anticipate some of the environmental factors the preceding generation had encountered. Through natural selection, those organisms which display epigenetic marks which allow them to anticipate environmental factors will have a selective advantage. However, this works only if correlations in the environment remain stable.

Associative learning-based model acquisition results from conditioning the organism to a stimulus–response relation. Reinforcement during an individual's lifetime generates an anticipative model which can quickly adapt to changes in environmental correlations, i.e. the model can be updated rapidly. Natural selection does not play a role in associative learning. There is, however, a special case through which it may affect future generations.

The Baldwin effect-based model acquisition is a result of natural selection. Unlike mutation and epigenetic-based models, though, the inheritable traits of the Baldwin effect include learning and behavioural changes. The idea is that individuals who display a certain noninheritable response to environmental stimuli which is of selective advantage will thrive and produce more offspring. Over many generations, the response is no longer environmentally induced, as the correlational structure becomes genetically embedded and thus heritable [56]. Again, it is of advantage only if correlations in the environment remain stable [57].

Against the background of these four ways of anticipatory model acquisition, Sims endorses the view that anticipatory models are a "ubiquitous feature of life" [58].

All four types of model acquisition (mutation-based, epigenetic inheritance-based, associative learning-based and Baldwin effect-based (if it exists)) produce internal models and would thus not be classified as local strong anticipation (anticipatory synchronization) as defined by Stepp and Turvey. Neither are these four types examples of delay compensation and therefore do not meet my four requirements for strong anticipation in extended obserpants.

4.6 Predictive Homeostasis

Biological systems maintain and recalibrate their function within a changing environment by means of two types of control systems: feedback and feedforward control:

> In a feedback control system, the output is quantitatively compared to the input to generate a measure of error. This error signal is then filtered out by a feedback controller that adjusts the input going back into the system. Feedback control represents a closed-loop design that mitigates error after it occurs, much in the same way that traditional models of homeostasis operate. Alternatively, feedforward control systems are open designs with controllers that adjust for error before the input enters the system. In this setup, environmental error is predicted or predetermined and applied to the input before it enters the systems to avoid or reduce system error ... [59]

Deans stresses the fact that feed-forward control and anticipatory processes are related in that both depend on correctly anticipating and adapting to future states and rely on predictions about future states to avoid or minimize error.

The classical concept of homeostasis used to be defined as a reactive process in a feedback loop, which is activated only after error occurs. Error was defined as any aberration from a set point, which was corrected by a central controller after environmental disturbance. However, the concepts of a set point and a central controller have been abandoned for systems which display anticipative behaviour:

> The reactive nature of homeostasis was questioned by Moore-Ede [in 1986], after noticing that the diurnal cycles of many physiological processes, including body temperature, plasma cortisol levels, growth hormone levels, and urinary potassium excretion, began before changes in light cues occurred and/or ahead of the sleep–wake cycles that were

> assumed to be regulating them (...) This led to the conclusion that these traits were regulated by a form of predictive homeostasis, a process that was more efficient because it minimized error. [60]

An example of predictive homeostasis is human thermoregulation, which is based on a predictive control system rather than a reactive one, mediated through internal models of the environment. If environmental change is stable and thus predictable, it provides fertile grounds for predictive control. By contrast, highly fluctuating and therefore unpredictable environmental change is steered through feedback control, i.e. a reactive response. Reactive control manifests itself in stress reactions, such as an increase in core body temperature and a decrease in peripheral temperature.

The body temperature of human beings and other large mammals differs between the peripheral temperature, which is measured in the skin, and the core temperature as measured in the internal organs. The core temperature changes more slowly than the peripheral temperature, which is exposed to environmental fluctuations. As endotherms, we regulate our body temperature through our metabolism within a narrow range. In extreme conditions, we have to exchange heat with our environment through radiation, evaporation, convection and conduction [61]. IJzerman describes penguins' huddling behaviour, which is steered by thermoregulation, as anticipative behaviour and a selection effect:

> Penguin huddling is a large-scale operation, not a last-minute emergency move. It requires each penguin to behave in a social way to the extent of surrounding itself with other penguins. Each bird is in a position of predicting its future body temperature based on its current social capital — the presence of other penguins. In effect, the penguin creates a weather report, a social weather report, and a prediction of its own body temperature based on both the weather report and the social weather report. How early penguins developed this predictive skill for the purposes of long-term thermoregulation we don't know. What we do know is that evolution selected for the trait of accurately predicting the presence of others. [62]

Social thermoregulation is an example of anticipatory behaviour which is not limited to penguins. In fact, the concept of social thermoregulation (as described in Section 3.3 as an example of obserpant reduction and expansion), which was introduced by IJzerman *et al.*, applies also to human beings [63].

IJzerman's main focus rests on the fact that, apart from behavioural thermoregulation, humans develop predictive models of themselves and their social environment. Physical reactions such as the constriction of blood vessels and piloerection (goosebumps) can be brought about not only by environmental temperature change but also through psychological conditions.

An example is the observation that a drop in skin temperature can be observed in infants if their mother leaves the room or a stranger is present [64]. IJzerman *et al.* conclude that thermoregulation is linked to attachment: Secure attachment through early thermoregulation provides a predictable environment, which enables adults to profit from predictive models formed when they were infants. Physical warmth is equated with social warmth and vice versa if mothers provided warmth through skin contact with the infant. As adults, this stable internal model of a close relationship enables them to compensate for the negative effects of being (briefly) socially excluded, for instance, by means of providing environmental warmth in the form of a cup of tea [65]. A sense of secure attachment which results from an early experience of caring warmth through skin contact provides a predictive social environment. Against this background, a predictive internal model which equates physical warmth with social warmth works bidirectionally.

As outlined in Section 3.3, social thermoregulation is an example of local strong anticipation in extended obserpants. The following section looks at a different kind of anticipative human interaction, which exhibits local strong anticipation in the form of anticipatory synchronization.

4.7 Anticipatory Synchronization in Human Interaction

Research on both biological and nonbiological systems has shown that small temporal feedback delays enhance a system's ability to anticipate unforeseen events. This includes anticipation in human physiology and interaction and is true in particular for nonroutine processes, i.e. processes or structures which exhibit chaotic, i.e. nonlinear behaviour: Coupled systems can anticipate chaotic behaviour if a small feedback delay is inserted [66].

Building on the work of Dubois, Stepp and Turvey, as well as the findings of Didier Delignières and Vivien Marmelat, who specialize in motor control and fractal physiology [67], psychologist Auriel Washburn *et al.* found that short visual-motor delays enhance interpersonal anticipatory synchronization. Whereas the above-mentioned examples, such as the anticipatory synchronization of lasers, are based on slave systems which are unidirectionally coupled to a master system, Washburn *et al.* also looked at bidirectional coupling [68]. This occurred when the master system itself was enslaved as a result of registering the slave's behaviour. Their motivation was the fact that

> … if anticipatory synchronization is to be useful in understanding complex, interpersonal coordination, then it must still occur in the context of bidirectional coupling between co-actors, and necessarily when both actors are able to see the other's behaviour. [69]

As they were dealing with human beings, the researchers used the term "producer" instead of "master" and "coordinator" instead of "slave". In their experiment, a coordinator was asked to synchronize his or her movements with those of a producer, whose chaotic movements were registered by a motion sensor attached to two fingers [70]. To initiate mutual enslavement through bidirectional coupling, the researchers instructed the producer and the coordinator to observe each other's movements on screens in two settings:

> The first, congruent, visual condition was designed so that both individuals had the same information about the coordinator's behaviour; the producer saw the coordinator's movements at the same perceptual delay that the coordinator experienced. In the second, incongruent, condition, the producer always viewed the coordinator's movements in real time while the coordinator saw his or her own movements with a feedback delay. [71]

The unidirectional setup showed the expected anticipatory synchronization of the coordinator with the producer (slave with master):

> … in the 400ms feedback delay condition, the movements of the coordinator began to lead those of the producer, indicating that the coordinator was in fact anticipating the producer's chaotic (…) movements. [72]

However, as the researchers observed anticipatory regulation also in bidirectional coupling, they concluded that it is not necessary for one communication partner to lead permanently — the roles of producer and coordinator can switch without compromising their anticipatory behaviour. In each case, feedback delays of 200–400 ms produced facilitated anticipatory behaviour, after a 600 ms delay, coordination broke down. The authors remark that this time span, which they observed in local behavioural anticipation in their experiments, is typical of delays in the human sensorimotor system and may be the basis for stable motor coordination. They emphasize, however, that the artificially induced delays in their experiment would be superimposed on the delays already inherent in the sensory-motor system. Thus, their results exaggerate natural anticipative processes [73].

Washburn *et al.* managed to describe a systemic whole composed of two coupled endo-systems. What makes their work unique is the bidirectional coupling of spatio-temporally extended obserpants who compensate for each other's delays. Each of the coupled obserpants exhibits local strong anticipation in the form of anticipatory synchronization.

Since temporal obserpant extensions as a result of coupling are boundary shifts towards the outside, the accompanying delay compensation extends our simultaneity horizon. This simultaneity horizon not only defines the systemic whole but also shapes the structure of our temporal interfaces — our Now.

The only reason why we can compensate for delays and thus anticipate future percepts is the fact that our Now is not just a point which separates the past from the future but has an extension and a structure. The following chapter presents a model of time suitable for the description of anticipative systems. It deals with the Now's nested internal structure which renders possible our perception of and interaction with multi-level, nested processes.

References

[1] In this context, biologist Eric Bittman correctly pointed out that entrainment is not synchronization, as synchronization does not require an oscillator. Sharing a common time period may be synchronous, but does not entail entrainment: "... synchronization is at most a narrow example of entrainment, if it is even that." (E. L. Bittman, Entrainment is NOT synchronization: An important distinction and its implications. *Journal of Biological Rhythms*, XX, 4, 2020.)

[2] S. Vrobel, *Fractal Time*. The Institute for Advanced Interdisciplinary Research, Houston, 1998.

[3] R. Nijhawan, Visual prediction: Psychophysics and neurophysiology of compensation for time delays. *Behavioural and Brain Sciences* (Cambridge University Press), 31, 197, 2008.

[4] *ibid*, p. 198.

[5] D. M. Eagleman, Prediction and postdiction: Two frameworks with the goal of delay compensation. In: Commentary section of R. Nijhawan: Visual prediction: Psychophysics and neurophysiology of compensation for time delays. *Behavioural and Brain Sciences* (Cambridge University Press), 31, 205, 2008.

[6] D. Eagleman. www.edge.org/conversation/brain-time. The Edge Foundation, 2022. Accessed 2.12.22.

[7] *ibid*.

[8] *ibid*.

[9] *ibid*.

[10] M. A. Changizi, Perceiving the present'as a framework for ecological explanations of the misperception of projected angle and angular size. *Perception*, 30, 195–208, 2001.

[11] D. A. Vaughn and D. M. Eagleman, Spatial warping by oriented line detectors can counteract neural delays. *Frontiers in Psychology Section on Consciousness Research*, 4, 1 November 2013.

[12] *ibid*, p. 1.

[13] Figure and caption: D. A. Vaughn and D. M. Eagleman, Spatial warping by oriented line detectors can counteract neural delays. *Frontiers in Psychology*

Section on Consciousness Research, 4, 2, 01 November 2013, Fig. 1, CC BY 3.0.

[14] *ibid*, p. 2.

[15] Figure and caption: *ibid*, p. 3; inset of Fig. 2, CC BY 3.0.

[16] D. Eagleman. www.edge.org/conversation/brain-time. The Edge Foundation, 2022. Accessed 2.12.22.

[17] C. Stetson, X. Cui, P. R. Montague and D. M. Eagleman, Illusory reversal of action and effect. *Journal of Vision*, 5, 769a, 651, 2006.

[18] T. Metzinger, *Being No One: Consciousness, the Phenomenal Self and the First-Person Perspective*. Foerster Lectures on the Immortality of the Soul, The UC Berkeley Graduate Council, 2006.

[19] *ibid* and personal correspondence, 2009.

[20] T. Metzinger, *Being No One*, MIT Press, Cambridge, MA, 2003, pp. 169–170.

[21] D. Eagleman. www.edge.org/conversation/brain-time. The Edge Foundation, 2022. Accessed 2.12.22.

[22] E. Pöppel, *Grenzen des Bewußtseins — Wie kommen wir zur Zeit und wie entsteht die Wirklichkeit*? Insel Taschenbuch, Frankfurt, 2000, pp. 38–42.

[23] N. Stepp and M. T. Turvey, On strong anticipation. *Cognitive Systems Research*, 11, 2, 1 June 2010.

[24] C. Huygens, *Letters to de Sluse, Letters*; no. 1333 of 24 February 1665, no. 1335 of 26 February 1665, no. 1345 of 6 March 1665, Societe Hollandaise Des Sciences, Martinus Nijho, 1895.

[25] H. Haken, *Synergetics, An Introduction*, 2nd edition, Springer, New York, 1978, p. 14.

[26] e.g. D. M. Dubois, Theory of incursive synchronization and application to the anticipation of delayed linear and nonlinear systems. *AIP Conference Proceedings*, Vol. 627, The American Institute of Physics, 2002, pp. 182–195.

[27] H. U. Voss, Anticipating chaotic synchronization. *Physical Review E*, 61(5), 5118, 2000.

[28] *ibid.*

[29] M. T. Turvey, *Lectures on Perception. An Ecological Perspective*, Routledge, New York, 2019, p. 414.

[30] *ibid*; H. U. Voss, Anticipating chaotic synchronization. *Physical Review E*, 61, 5115, 2000.

[31] N. Stepp and M. T. Turvey, On strong anticipation. *Cognitive Systems Research,* 11(2), 137–138, 1 June 2010.

[32] M. T. Turvey, *Lectures on Perception. An Ecological Perspective*, Routledge, New York, 2019, p. 416.
[33] C. Deans, Biological prescience: The role of anticipation in organismal processes. *Frontiers in Physiology*, 12, 672457, 2, 2021.
[34] I. P. Pavlov, *The Work of the Digestive Glands*, Charles Griffin and Company, Ltd., London, 1902.
[35] H. Lefebre, *Rhythmanalysis: Space, Time and Everyday Life*, Éditions Syllepse, Paris, 1992. English translation: Continuum, London, 2004.
[36] C. Deans, Biological prescience: The role of anticipation in organismal processes. *Frontiers in Physiology*, 12, 672457, 3, 2021.
[37] *ibid*, p. 4.
[38] *ibid.*
[39] Reprinted from C. Deans, Biological prescience: The role of anticipation in organismal processes. *Frontiers in Physiology*, 12, 4, December 2021. Article 672457 (CC BY 4.0). www.frontiersin.org.
[40] A. Boussard, A. Fessel, C. Oettmeier, L. Briard, H.-G. Döbereiner and A. Dussutour, Adaptive behaviour and learning in slime moulds: The role of oscillations. *Philosophical Transactions of the Royal Society B*, 376, 20190757, 5.
[41] *ibid*, p. 4.
[42] *ibid*, p. 7.
[43] *ibid*; K. Alim, N. Andrew, A. Pringle and M. P. Brenner, Mechanism of signal propagation in Physarum polycephalum. *Proceedings of the National Academy of Sciences USA*, 114, 5136–5141, 2017.
[44] A. Boussard, A. Fessel, C. Oettmeier, L. Briard, H.-G. Döbereiner and A. Dussutour, Adaptive behaviour and learning in slime moulds: The role of oscillations. *Philosophical Transactions of the Royal Society B*, 376, 20190757, 8.
[45] T. Saigusa, A. Tero, T. Nakagaki and Y. Kuramoto, Amoebae anticipate periodic events. *Physical Review Letters*, 100, 18101ff, 2008.
[46] *ibid*, p. 18101-1.
[47] O. Falik *et al.*, Rumour has it ...: Relay communication of stress cues in plants. *PLoS One*, 6(11), November 2011, paper e23625; R. Karban *et al.*, Herbivore damage to sagebrush induces resistance in wild tobacco: Evidence for eavesdropping between plants. *OIKOS*, 100, 325–332, 2003.
[48] R. Karban *et al.*, Herbivore damage to sagebrush induces resistance in wild tobacco: Evidence for eavesdropping between plants. *OIKOS*, 100, 326, 2003.

[49] *ibid*, p. 331.

[50] C. Dean, Biological prescience: The role of anticipation in organismal processes. *Frontiers in Physiology Section on Integrative Physiology*, 12, 11, 17 December 2021.

[51] DNA methylation is "the addition or removal of a methyl group (CH3), predominantly where cytosine bases occur consecutively. DNA methylation was first confirmed to occur in human cancer in 1983, and has since been observed in many other illnesses and health conditions." (B. Weinhold, *Environmental Health Perspectives*, 114(3), A 163, March 2006.)

[52] C. Deans, Epigenetic processes as anticipatory mechanisms: Insect polyphenism as an exemplar. In: M. Nadin (ed.) *Epigenetics and Anticipation, Cognitive Systems Monographs*, Vol. 45, Springer Nature, Switzerland, 2022, p. 122.

[53] M. Sobral, L. Sampedro, I. Neylan, D. Siemens and R. Dirzo, Phenotypic plasticity in plant defense across life stages: Inducibility, transgenerational induction, and transgenerational priming in wild radish. *PNAS*, 118(33) e2005865118, 1, 13 August 2021.

[54] M. Sims, Many paths to anticipatory behavior: Anticipatory model acquisition across phylogenetic and ontogenetic timescales. Biological Theory, 1–20, 2023.

[55] *ibid*, p. 2.

[56] *ibid*, p. 16.

[57] As Sims concedes, there are no examples which directly support the existence of the Baldwin effect. However, there are promising research results (F. Mery and T. J. Kaweki, Experimental evolution of learning ability in fruit flies. *Proceedings of the National Academy of Sciences of USA*, 99, 14274–14279, 2002; and M. Sims, Many paths to anticipatory behavior: Anticipatory model acquisition across phylogenetic and ontogenetic timescales. *Biological Theory*, 15, 2023).

[58] M. Sims, Many paths to anticipatory behavior: Anticipatory model acquisition across phylogenetic and ontogenetic timescales. *Biological Theory*, 18, 2023.

[59] C. Deans, Biological prescience: The role of anticipation in organismal processes. *Frontiers in Physiology Section on Integrative Physiology*, 12, 14, 17 December 2021.

[60] *ibid*, p. 5; M. C. Moore-Ede, Physiology of the circadian timing system: Predictive versus reactive homeostasis. *American Journal of Physiology*, 250, 735–752, 1986.

[61] Homeostasis — Thermoregulation — Biology Libre Texts. https://bio.libretexts.org/Bookshelves/Introductory_and_General_Biology/General_Biology_1e_(OpenStax)/7%3A_Animal_Structure_and_Function/33%3A_The_Animal_Body_-_Basic_Form_and_Function/33.3%3A_Homeostasis.

[62] H. R. IJzerman, *Heartwarming. How Our Inner Thermostat Made Us Human.* W.W. Norton & Company, London, 2021, p. 88.

[63] H. R. IJzerman H., J. A. Coan, F. M. A. Wagemans, M. A. Missler, I. van Beest, S. Lindenberg and M. Tops, A theory of social thermoregulation in human primates. *Hypothesis and Theory, Frontiers in Psychology*, 6, 21 April 2015 (online article).

[64] M. D. S. Ainsworth and S. M. Bell, Attachment, exploration, and separation: Illustrated by the behavior of one-year-olds in a strange situation. *Child Development*, 41, 49–67, 1970.

[65] H. R. IJzerman, M. Gallucci, W. T. J. L. Pouw, S. C. Weiβgerber, N. J. Van Doesum, K. D. Williams, Cold-blooded loneliness: Social exclusion leads to lower skin temperatures. *Acta Psychologica*, 140, 283–288, 2012.

[66] D. M. Dubois, Computing anticipatory systems with incursion and hyperincursion. In: D. M. Dubois (ed.) *Computing Anticipatory Systems: CASYS — First International Conference. AIP Conference Proceedings*, Vol. 437, American Institute of Physics, Melville, NY, 1998; A. Washburn, R. W. Kallen, C. A. Coey, K. Shockley and M. J. Richardson, Interpersonal anticipatory synchronization: The facilitating role of short visual-motor feedback delays. *Proceedings of the 37th Annual Meeting of the Cognitive Science Society*, 2015.

[67] N. Stepp and M. T. Turvey, On strong anticipation. *Cognitive Systems Research*, 11(2), 158, 1 June 2010; D. Delignières and V. Marmelat, Strong anticipation and long-range cross-correlation: Application of detrended cross-correlation analysis to human behavioural data. *Physica A*, 394, 47, 47–60, 2014.

[68] A. Washburn, R. W. Kallen, C. A. Coey, K. Shockley and M. J. Richardson, Interpersonal anticipatory synchronization: The facilitating role of short visual-motor feedback delays. In: *Proceedings of 37th Annual Meeting of the Cognitive Science Society*, 2619, 2015.

[69] *ibid*, p. 2621.

[70] "Chaotic" here refers to unpredictable behaviour which strongly depends on initial conditions and manifests, among other parameters, as a positive Lyapunov exponent.

[71] *ibid*, pp. 2620–2621.

[72] *ibid*, p. 2621.

[73] *ibid*, p. 2623; see also: S. Wallot and G. van Orden: Ultrafast cognition. *Journal of Consciousness Studies*, 19(5–6), 141–160, 2012.

Chapter 5

A Model of Time to Describe Anticipative Systems

Manifestations of strong anticipation are local or global, model-based or model-free, include or exclude representations, are based on compensated delay and distance or on the coordination of long-term correlations. Despite this variety, all are based on underlying assumptions about the nature and texture of the world. This chapter focuses on one such assumption: our presuppositions about the nature and structure of time.

As late and living researchers have pointed out, our theories and models of the world are secondary products based on our primary perception of time, which we experience as duration, simultaneity and succession [1]. Rosen differentiated between simultaneity and precedence (whose temporal extension corresponds to successive order). He states that simultaneity

> ... will specify how the values assumed by particular observables in the system at an instant of time are related to the value assumed by the other observables,

whereas succession

> will specify how the values assumed by particular observables at a given instant are related to the values assumed by these or other observables at other instants. [2]

The distinction between succession and simultaneity is essential when we wish to describe any dynamical system. For strong anticipative models, further structural constraints need to be taken into account, such as the internal structure of our Now (or other reference systems) and the spatio-temporal embedding of rhythms when new systemic wholes are formed. My Theory of Fractal Time provides a suitable basis to describe anticipatory systems [3]. Delay compensation and synchronization become formally describable in terms of the length, the depth and the density of time [4].

5.1 Nested Rhythms, Fractal Nows

Delays implicitly presuppose duration, simultaneity and succession: A delay spans a certain amount of time in which successive events create a before–after relation. Local strong anticipation in extended obserpants increases simultaneity by adding external rhythms to internal ones. Anticipatory synchronization is generated through the coupled master–slave relation, which continuously reduces the gap between $x(t+1)$ and $x(t)$ until succession has turned into simultaneity (i.e. until master and slave are synchronized).

But not only coupled systems with obvious delays display a nested temporal structure consisting of both succession and simultaneity. Coupled systems without delays presuppose the same temporal extensions: Succession and nested simultaneity are generated by the internal dynamics of an agent, say a biological entity, because processes on all levels, from organism to organ to cells, happen simultaneously. Cell divisions, for instance, create succession, as they generate before-and-after relations. At the same time, they create nested simultaneity because mitosis might happen while melatonin is secreted, a late dinner is digested and our blood pressure and body temperature slowly drop in anticipation of a night's sleep. We are embedded into and embed temporal rhythms of varying lengths. Our embedding environment provides yet more degrees of simultaneity: astronomical cycles, tidal rhythms, circadian rhythms and others of which we are not aware. The same is happening at the other end of the scale, in neural oscillations in our brain [5].

From the obserpant's perspective, such internal and external processes provide a nested background against which he or she structures percepts in the Now — our interface which connects us to and separates us from the world.

When rhythms of differing lengths overlap and/or are embedded in rhythms with more extended phases, the nested structure of the Now emerges. It manifests itself in the complex communication between cells, organs, the entire body and our physical and social environment. It is this nesting of internal and external rhythms which creates a temporally extended obserpant. Nadin stresses that one clock suffices to measure successive processes, but living systems require a more differentiated time setting on several levels:

> … it has been repeatedly demonstrated that an anticipatory system has at least two clocks, i.e., correlated processes unfolding at different time scales (…). At various levels of the living, several clocks are at work: some very fast (at nanosecond speed); others in the domain of the 'gravitational' clock; and yet others are very slow. Therefore, Rosens's model unfolding at faster than real time is probably a distributed anticipatory process with many models operating at various time scales. [6]

The idea of such nested clocks goes back to phenomenologist Edmund Husserl who, in 1928, first provided a model which — like Rosen's model half a century later — described the obserpant's present as a result of both past and future events [7].

Husserl's phenomenological approach focused on the nested structure of the Now with its anticipatory and postdictive characteristics. He saw the obserpant's Now not as a point which separates the past from the future but as an extended temporal window with a nested internal structure. This conviction was based on his observation that when we listen to a piece of music, we do not just hear an uncorrelated succession of individual notes — as we would if the Now were point-like, without extension — but a tune.

According to Husserl, this is possible because the listener remembers, within his consciousness of the present, the note which was just played and still lingers on and anticipates the notes which will follow the presently sounding one. He denotes the lingering of the note just played as *retension* and the expectation that another note would follow as *protension*. Retension is not simply a memory but a lingering memory within the consciousness of the present, i.e. it has a temporal extension. Likewise, protension is not simply an expectation but, as an extension of the Now into the future, a form of strong anticipation.

Within the consciousness of the present, a cascade of nested retensions and protensions structure and span the Now, embedding and overlapping remembered and anticipated events. Each Now hosts retensions and protensions of the preceding Nows, which continue to sink into the past with each new Now, each new contextualization. In this cascade, each new embedding modifies the preceding retensions and protensions, adapting their meaning to the new context (see Fig. 5.1).

French philosopher Henri Lefebvre coined the term "rhythmanalysis" to describe the relationship between internal (biological) and external (social and physical) rhythms as they manifest within the human body. He describes the internal and external polyrhythmic interactions as a nested structure:

> ... the living body can and must consider itself as an interaction of organs situated inside it, where each organ has its own rhythm but is

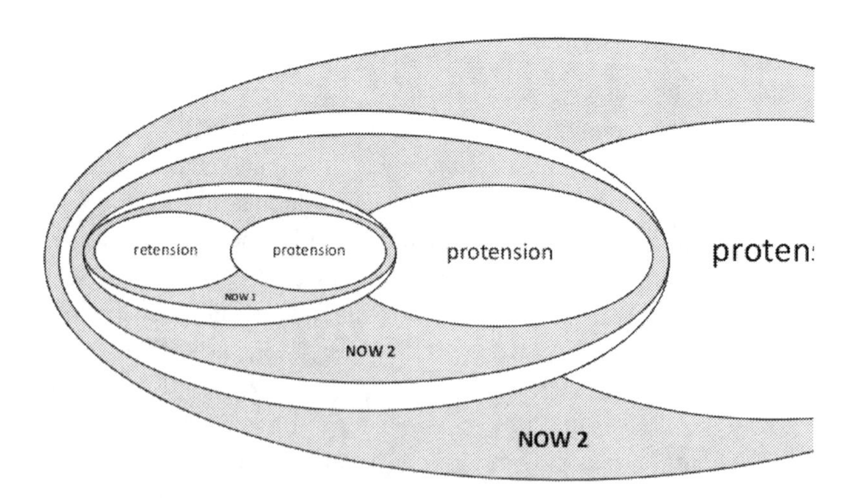

Fig. 5.1. Nested Nows: Protension and retension of Now 1 become retension in Now 2, etc.

> subjected to a spatio-temporal whole [globalité]. Furthermore, this human body is the site and place of interaction between the biological, the physiological (natural) and the social (often called the cultural), where each of these levels, each of these dimensions, has its own specificity, therefore, its space-time: its rhythm. Whence the inevitable shocks (stresses), disruptions and disturbances in this ensemble whose stability is absolutely never guaranteed. Whence the importance of scales, proportions and rhythms. [8]

Both the obserpant's structure of the nested Now as described by Husserl and Lefebvre's nested arrangement of rhythms within and between the human body and its environment can be measured in terms of the length, depth and density of time.

5.2 Fractal Time: Connecting Succession and Nested Simultaneity

In 1994, I developed my Theory of Fractal Time, which differentiates between two temporal dimensions and their relation [9]. It is based on the concept of a fractal as defined by mathematician Benoit Mandelbrot in the 1970s, building on Lewis Fry Richardson's work [10]. In analogy to spatial fractals, I relate the number of similar elements on all levels in a nested structure with the scaling factor. However, unlike Mandelbrot's notion of fractal time, which plotted, for instance, variations in cotton prices from an exo-perspective, my concept relates succession and nested simultaneity as two temporal dimensions from an endo-perspective, i.e. within an obserpant frame:

- Δt_{length} is the number of incompatible events. It defines the temporal dimension of succession and, in Husserl's musical example, it stands for successive, uncorrelated notes within the obserpant's Now.
- Δt_{depth} is the number of compatible events. It defines the temporal dimension of nested simultaneity. In the musical example, it stands for overlapping or nested notes within the obserpant's Now.
- $\Delta t_{\text{density}}$ is the fractal dimension calculated from the relation between Δt_{length} and Δt_{depth}. It is a measure to compare degrees of temporal

complexity. In the musical example, it measures the internal temporal complexity of a composition [11].

The resulting nested structure, composed of parallel, overlapping or embedded percepts or events of various lengths, creates a temporal fractal structure. A fractal, be it of a spatial or temporal nature, is a structure which exhibits detail on several scales. Mathematical fractals such as the Mandelbrot set exhibit an infinite number of nested scales, which all reveal more and more detail as one zooms into the structure. Natural fractals, by contrast, have an upper and a lower limit to the number of nested scales.

The literature often lists self-similarity (scaling behaviour) as a defining feature of fractals, meaning that the whole consists of smaller copies of itself, nested in a number of scales. While this is often the case, most fractal structures display merely a statistical self-similarity. In fact, as we shall see in Barnsley's approach (in the following), self-similarity is not a necessary condition for a structure to qualify as a fractal.

Spatial structures such as the Koch curve (see Fig. 5.2) are artificial shapes constructed by a generation rule: Take the mid-third of a line and

Fig. 5.2. The self-similar Koch curve [13].

replace it with an equilateral triangle with side lengths of one-third of the original line. Then repeat this procedure for each of the line sections. Continue this generation rule *ad infinitum* and you have a Koch curve.

The fractal dimension of such self-similar structures (also called the similarity dimension) can be calculated as $D = \log(n)/\log(s)$, with n representing the number of self-similar structures and s the scaling factor. For the Koch curve, the fractal dimension is $D = \log 4/\log 3 = 1.2618$. The general definition of a fractal is a structure with a broken dimension, as opposed to a Euclidean shape like a square. A square which consists of, say, 9 smaller squares of 1/3 of its size has the fractal dimension $D = \log 9/\log 3 = 2$. This shows it is not a fractal, as its fractal dimension is an integer [12].

However, fractals in nature are rarely constructed with a generation rule an obserpant might deduce. Therefore, a more general method of measuring the fractal dimension of a structure was introduced by mathematician Michal Barnsley [14]. He stressed that a cube of integer dimension 2 would also be defined as a fractal if studied on more than one scale. In fact, any structure which can be described in terms of its scaling properties would be fractal.

Barnsley measures the fractal dimension D of a structure in a Euclidean metric space by covering that structure with closed just-touching boxes of side length $(1/2^n)$: If $D = \lim n \to \infty \{\ln (N_n (A))/\ln (2^n)\}$, then A has the fractal dimension D (where N_n (A) denotes the number of boxes of side length $(1/2^n)$ which intersect the attractor, that is to say, the structure in this space which is covered by boxes (see Fig. 5.3) [15].

By shifting fractality into the eye of the measuring obserpant, Barnsley made the obserpant the defining instance: Fractality depends on the obserpant's internal complexity, i.e. how many different scales he or she can make out in their environment and how complex is their range of responses [16]. Barnsley's box-counting method can be applied to both self-similar and nonscaling structures. Figure 5.3 depicts the first three stages of determining the fractal dimension of the Koch curve by means of the box-counting method.

Fractal time series as those presented by Mandelbrot, like variations in the prices of cotton, are correlations between different time scales which display statistical self-similarity. In such time series, data are structured in retrospect from an exo-perspective.

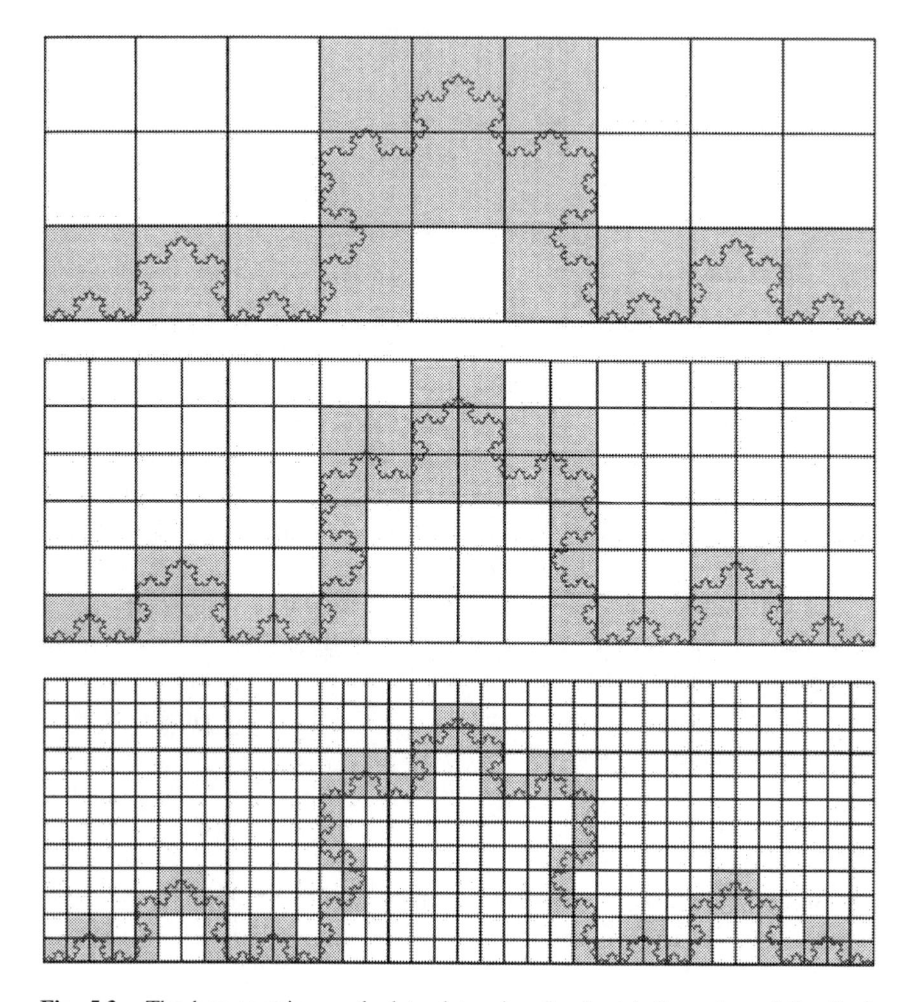

Fig. 5.3. The box-counting method to determine the fractal dimension of the Koch curve [17].

By contrast, strong anticipation in fractal time happens and is registered in real time, not in retrospect (as in time series analysis). The coordination between internal and external fractal structures occurs at the interface between the obserpant and environment. In the obserpant's nested Now, the recurring structures on the various nested levels are perceived simultaneously.

A common temporal fractal is $1/f$ scaling, also known as "pink noise" or "flicker noise", which human beings can perceive in real time.

Its amplitude decreases with increasing frequency, i.e. all frequencies have roughly the same volume. While it is ubiquitous in nature, including the dynamics of human beings, the origin of $1/f$ (pink) noise is not known [18].

Other fractal colours of noise include white and brown noise. White noise is the simplest scaling noise. It displays short-scale fluctuations, has no correlations in time and is thus unpredictable. Pink noise exhibits more long-scale fluctuations than white noise and is therefore more predictable. Brownian noise, which is characterized by long-scale fluctuations, is even more correlated and thus more predictable than white or pink noise (see Fig. 5.4).

Mandelbrot first recognized the occurrence of $1/f$ noise in nature, as, for instance, the $1/f$ fluctuation of the annual flood levels of the river Nile [19].

Mathematician Richard F. Voss compared white, brown and pink noise in terms of their spectral density $S(f)$ (also known as power spectrum). White noise has a spectral density of $1/f^0$, pink noise of $1/f^\beta$ (with $\beta \approx 1$) and Brownian noise of $1/f^2$ [20]. He drew attention to the ubiquity of $1/f$ noise in all walks of life, from processes within human beings to those in the environment and even in symbolic sequences:

> Many natural time series look (…) like the 1/f noise or *pink noise* (…). A wide variety of measured quantities, from electronic voltages and time standards to meteorological, biological, traffic, economic and musical quantities, show measured $S(f)$ varying as $1/f^\beta$, with $\beta \approx 1$ over many decades. (…) In most cases, the physical reason for this long-range power-law behaviour remains a mystery. [22]

Voss stresses that a certain degree of complexity is necessary to generate $1/f$ noise and that it appears to be related to human communication, such as language and music [23].

As we see in Chapter 11, pink noise has a positive effect on human health and well-being [24]. It is a statistical fractal, so its structure can usually only be detected in retrospect, at the end of an extensive period of time. However, the temporal fractal $1/f$ (pink) noise is perceived in real time within the obserpant's Now, as Kelty-Stephen *et al.*'s experiment on long-range correlations in synchronization showed (see Section 2.6.2).

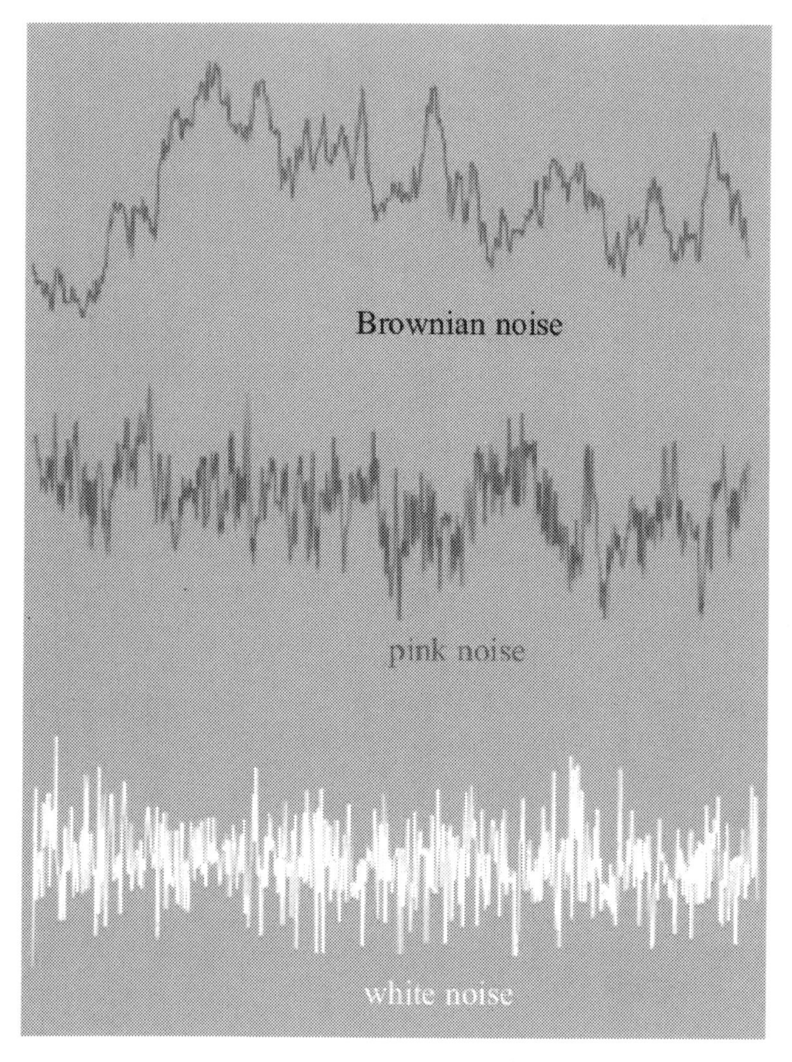

Fig. 5.4. Brown, pink and white noise [21].

Global strong anticipation as a result of the coordination of long-term correlations supports the observation that our Now is remarkably stretchable and has a scaling structure to host the successive and simultaneous extensions, i.e. Δt_{length} and Δt_{depth}, of statistical fractals, such as $1/f$ noise.

5.3 Fractal Spacetime

Whether we talk about my Theory of Fractal Time as generated by obserpant perspectives or inherent in our environment, the principle of scale relativity as developed by French astrophysicist Laurent Nottale rules this new fractal way of modelling reality [25]. Nottale's Theory of Scale Relativity extends Einstein's principle of relativity to scale transformations. The geometry of this continuous but nondifferentiable fractal spacetime is resolution-dependent. The novelty of Nottale's approach is that he applies the notion of fractality to the geometry of spacetime itself. It aims at describing a universal scale-relativistic physics in which the equations of physics are invariant under scale transformations. The principle of scale co-variance is universal and applicable, for instance, to biological, inorganic, social and economic phenomena. It allows for evolutionary analogies between phylogeny and ontogeny on the basis that both may manifest, at each scale of organization, an underlying memory phenomenon [26]. On the one hand, Nottale attributes a fractal structure to spacetime and, on the other hand, he defines scaling relativity as being resolution-dependent. This is reminiscent of Barnsley's box-counting approach, as it also places fractality into the obserpant's spatio-temporal perspective.

Together with psychiatrist Pierre Timar, Nottale describes temporal structures from an endophysical perspective, which shifts the scale-relativistic character of our perception of time into the focus of their approach [27]. The researchers focus on temporal incommensurability between obserpants and their environments (including other obserpants) and disorders resulting from such temporal discord.

Philosopher Kerry Welch has managed to reconcile my Theory of Fractal Time with Nottale's fractal spacetime, which means that she combined the fractal perspective of an endo-obserpant with a theory of fractal spacetime based on scaling relativity [28]. It should be noted, however, that Nottale's approach is also of an endophysical nature.

Another fractal description of spacetime was developed by mathematician and physicist Mohammed El Naschie [29]. The fractal structure in his Cantorian spacetime is also inherent in spacetime itself. The geometry is based on a fractal structure whose limit is a plane-filling Peano-like curve.

A Peano–Moore curve has also been suggested by Ord as an underlying structure of fractal spacetime [30]. Cantorian spacetime re-interprets Young's double-slit experiment by claiming that the wave interference is a result of the geodesic waves of spacetime itself. The underlying discrete hierarchical fractal spacetime has an infinite number of dimensions. This model, which is known as El Naschie's e-infinity, aims at the description of both relativity and quantum particle physics [31].

Time series analysis has seen a vast number of fractal analytical tools [32]. One such method, flicker-noise spectroscopy, was developed by physicist Serge Timashev to measure time series on several scales simultaneously with a fractal measuring chain [33]. Timashev's method recognizes fractal structures in chaotic signals and reveals information which is hidden in correlation links. These are typically located in a sequence of irregularities such as spikes, jumps, and discontinuities in derivatives of different order.

Dubois' anticipatory systems (see Chapter 2) are another example of fractal structures, as they incursively (and hyperincursively) generate a model of themselves which they then map onto themselves [34].

The theories of fractal spacetime and methods briefly described above all have a common denominator: They are based on scaling structures, either as a property of the fabric of our universe, a fractal measuring method, or as a generation mechanism. They allow a bridging of nested scales. If we imagine the scale-independent recurring structure to become the measuring rod, as opposed to the level-dependent measure of distance or time, spacetime on all scales warps around the new constant (the recurring structure). Chapter 12 extrapolates on this constant which facilitates an anticipative perspective.

5.4 Long-Range Correlations: Coordination on Nonlocal Time Scales

In Chapter 2, we looked at the distinction between local and global strong anticipation as defined by Stephen and Dixon. They suggest that global strong anticipation arises from the coordination of internal and external long-range correlations, possibly resulting from the ubiquity of $1/f$ fractal scaling. The fact that many natural systems display fluctuations on nested time scales may hint at the possibility that global strong anticipation is also ubiquitous [35].

Marmelat *et al.* also suggested that matching long-range correlations of an organism with long-range correlations in its environment might be a strong indicator of anticipatory behaviour. In an experiment on interpersonal coordination, participants were asked to synchronize with a partner, each moving a pendulum. Although there was little local correlation in the dyad's pendulum movements, the researchers found a long-term correlation. This correlation was detected in the scaling properties of the time series the participants had generated with their synchronization attempts. Although the short-term behaviour was less correlated, a strong correlation between long-term fractal exponents was revealed [36].

If short-term coupling was apparently not responsible for the matching of long-term fluctuations, could the ubiquity of $1/f$ scaling be the underlying cause? A temporal fractal such as $1/f$ noise is certainly a promising candidate for an underlying facilitator of global strong anticipation. It may thus be an underlying mechanism for matching fluctuations within the obserpant with those within the obserpant's environment. As cognitive and biological processes within the obserpant exhibit $1/f$ scaling, coordination with external $1/f$ fluctuations seems likely.

Whether or not global patterns such as $1/f$ scaling may result from local coupling of oscillations is a matter of debate. Ian D. Colley and Roger T. Dean explain $1/f$-type fluctuations with short-range models including multiple delays. They used data from musical contexts but found that short-range processes and $1/f$ fluctuation co-exist also in gait, heartbeat and neural oscillations. The researchers suggest that $1/f$ fluctuation is possibly not a manifestation of long-range correlations [37]. However, a large section of the scientific community connects $1/f$ noise with long-range correlations and anticipation [38].

A nonlocal mechanism which connects long-term and short-term rhythms was presented by neuroscientist Simo Monto, who studied nested oscillations and nested synchrony in the brain (see Fig. 5.5). His account of global rhythms steering nested local ones is an example of global strong anticipation [39]:

> We have introduced and tested a novel cross-frequency interaction model of nested synchrony. In this model, the neuronal interareal

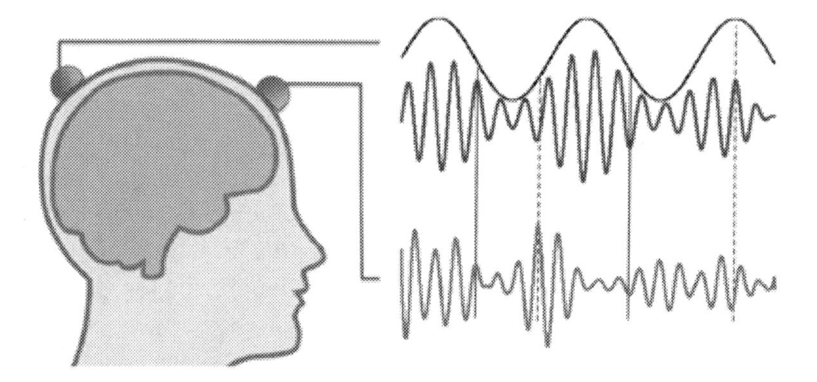

Fig. 5.5. Nested oscillations and nested synchrony: In nested oscillations, the phase of the slower oscillation modulates the higher-frequency amplitude measured at the same scalp location (two upper traces) but not that measured at a different location (lower trace). In contrast, nested synchrony means that the two faster oscillations become more tightly coupled in certain phases of the slow oscillation (around solid vertical lines) than in other phases (around dashed vertical lines) [41].

> oscillatory interactions are modulated by a lower-frequency oscillation, in an analogous fashion to nested oscillations discovered previously. [40]

In Monto's nonlocal cross-frequency interaction,

> the phase of the slower oscillation regulates inter-areal synchrony in the higher frequency band. [42]

This nonlocal mechanism binds neuronal behaviour across different areas of the brain and integrates global interactions across scales. In other words, it is an example of a fractal time series in which low-frequency oscillation steers nested neuronal interareal oscillatory interactions. This makes the synchrony in the higher frequency band a manifestation of global strong anticipation.

5.5 Long-Term Global Coordination: Correlating Fractal Dimensions

In Section 4.7, we saw in Washburn *et al.*'s experiment how human co-actors anticipate each other's chaotic, i.e. unpredictable, movements.

Anticipatory synchronization was established between two bidirectionally coupled co-actors with a delay time of 200–400 ms [43]. The researchers expanded their experiment in a second phase to determine whether, apart from the discovered short-term synchronization, long-term global coordination also exists between the co-actors.

As Delignières and Marmelat had shown, global coordination between co-actors exists in the form of long-term correlations. These manifest themselves in the co-actors' self-similar behavioural dynamics [44]. Washburn *et al.*'s research focuses on Delignières and Marmelat's result that each individual showed variability in their sensorimotor behaviour which encompassed recurrent structures over nested time scales. This variability could be quantified and thus compared:

> … this global coordination can be quantified by comparing the fractal (self-similar) structure of the behavioural variability found within each of the two actors' concurrent, coordinated behaviours. [45]

Washburn *et al.* measured the spatial self-similarity between the co-actors' movements (plotted in a 2-dimensional plane) as the correlation between their fractal dimensions. They found evidence of long-term complexity matching in a

> … very strong positive relationship between the fractal dimension of the producer movements and the fractal dimension of the associated coordinator movements. [46]

The researchers concluded that the coordinators had assimilated the global dynamics of the producer's chaotic movements. This assimilation occurred independently and was not correlated to the strength of local coordination revealed in their first experiments. The underlying fractal pattern was not based on the short-term coordination in the actors' behaviour but appeared to be a contextual underlying structure. Washburn *et al.*'s results are thus also an example of global strong anticipation. They stress that their findings underline their nonmodel approach, in which there is no need for

> a set of internal, 'black-box' compensatory neural simulations, representations, or feed-forward motor programs. [47]

Further evidence for global strong anticipation was presented by Marmelat *et al.* in an experiment which investigated the coordination of stride time dynamics [48]. It has been shown that stride variability with a relatively high fractal dimension is an indicator of health. Conversely, old age and a number of pathological conditions correlate with low fractal dimensions in human gait [49]. Low fractal dimensions also mean fewer long-term correlations. And as human beings display higher fractal dimensions in their gait variability than an isochronous pacesetter, the researchers decided to ask human beings to act as leaders rather than using a metronome. The participants were divided into pairs of leaders and followers, with the followers attempting to synchronize their gait with the leaders. It turned out that the human pacemakers influenced the fractal exponent α of the followers [50]:

> ...we did not find any statistical difference between α-leaders and α-followers. The agreement between $\alpha_{\text{long-term}}$ — exponents (...) implies that synchronization was mainly achieved by long-range processes and not primarily by local over-corrections. [51]

Most importantly, these results confirm the presence of global strong anticipation, as followers not only synchronized with the leaders but also anticipated their gait:

> [... the followers] did not react to the behaviour but actively anticipated the occurrence of the steps. This anticipation is important because we could have expected that stride time series of followers matched those of leaders with a constant lag + 1, if followers were only reacting to leaders. In this situation, α_{follower} and α_{leader} would have been the same, but only as a result of followers mimicking leaders. [52]

In other words, the followers had compensated for the delays, thus displaying strong anticipative behaviour. As α_{follower} and α_{leader} were statistically self-similar, the researchers concluded that, in nursing and rehab, human pacemakers are more efficient than isochronous ones (e.g. metronomes), especially for older people whose fractal exponents have decreased. The patients' fractal exponents may be increased by synchronizing their gait with a human follower, thus improving their gait variability, i.e. their range of possible responses.

One possible reason for the loss of variability in old age, as revealed in decreased fractal dynamics, is that older people or individuals with a neurodegenerative disease concentrate on gait step-by-step, successively adapting to, say, auditory stimuli [53]. In terms of my Theory of Fractal Time, adjusting synchronization locally, step-by-step, increases Δt_{length} (succession), while Δt_{depth} (nested simultaneity) decreases. In this case, internal complexity, i.e. the range of possible responses, which is measured in the fractal dimension $\Delta t_{\text{density}}$ (the density of time) also decreases.

5.6 Delay Compensation Turns Succession into Simultaneity

Δt_{depth} (nested simultaneity) is created when a new context requires us to adapt by shifting the interfacial cut between ourselves and our environment outwards, so as to incorporate and compensate for an environmental delay or distance. We become spatio-temporally extended obserpants who have moulded into a systemic whole with parts of our environment. This act of contextualization — that is to say, the generation of Δt_{depth} — is an acquired skill and possibly a selection effect. Merging interfaces and compensating delays decreases temporal complexity, which enables us to navigate the world more smoothly.

Whenever a delay is compensated, Δt_{length} is transformed into Δt_{depth}. What first belonged to the outer realm of the interfacial cut within the obserpant (neurocognitive delays) or between the obserpant and his or her environment (environmental delays) is now (after recalibration) part of a new systemic whole which includes the obserpant. And as the new systemic whole by definition has a common simultaneity horizon, which was recalibrated after the delay compensation, what was once succession (the delay) is transformed into simultaneity. The temporal dimension of Δt_{length} has been transformed into Δt_{depth}. And as we have identified delay compensation as a manifestation of local strong anticipation in obserpant extensions and in anticipatory synchronization, we may conclude that local strong anticipation is, in general, a result of transforming Δt_{length} into Δt_{depth}.

In the keypress experiment by Stetson *et al.* [54], subjects compensated for the inserted delay after a short period of recalibration. What was first an added successive temporal extension was turned into a simultaneous temporal extension as a result of delay compensation (i.e. anticipation). The subjects' Now transformed Δt_{length} into Δt_{depth}.

When the existing and thus compensated delay was shortened or removed, subjects perceived a reversal of temporal order. After recalibrating their temporal interface by adding an interval of Δt_{length} (succession), they were no longer confused, as internal and external temporal dimensions were congruent again. By adding an interval of Δt_{length} (succession), they compensated for the removed delay. At the same time, the subjects' temporal interface (his or her Now) would unwittingly remove a layer of Δt_{depth} (which was created during the first compensatory act when the delay compensation turned Δt_{length} into Δt_{depth}). This removal of a layer of Δt_{depth} would go unnoticed, as, in general, obserpants are not aware of the number of nestings, i.e. the degree of simultaneity in their Now.

As we have probably compensated many a delay or distance during our phylogenetic and ontogenetic evolution, I would venture to claim that there is a wealth of such compensatory acts we perform every day without being aware of it. Revealing them would be a rewarding undertaking and not just be of epistemological concern. Nonlocality may be unmasked as delay and distance compensation (more on this topic in Chapter 12).

To conclude this chapter, my Theory of Fractal Time provides, with its temporal dimensions Δt_{length}, Δt_{depth} and $\Delta t_{\text{density}}$, a model which is suitable for defining anticipative systems. Local strong anticipation occurs if the obserpant's temporal interface — the Now — transforms Δt_{length} into Δt_{depth}. Global strong anticipation shows itself in the coordination of long-term fractal correlations, i.e. the fractal dimension of $\Delta t_{\text{density}}$.

Both delay compensation in extended obserpants and in anticipatory synchronization transform succession into simultaneity (i.e. they remove succession and create simultaneity). Compensating a removed delay creates succession and removes simultaneity, which means that no anticipatory scenario can evolve. The generation or presence of Δt_{depth} is a presupposition for strong anticipation.

References

[1] R. Rosen, *Anticipatory Systems*, 2nd edition, Springer, 2012 (First published 1985 by Pergamon Press); E. Pöppel, Erlebte Zeit und Zeit überhaupt: Ein Versuch der Integration. In: H. Gumin, and H. Meier (eds.) *Die Zeit. Dauer und Augenblick.* Piper, Munich, 1989; E. Pöppel, *Grenzen des Bewußtseins — Wie kommen wir zur Zeit, und wie entsteht die Wirklichkeit?* Insel Taschenbuch, Frankfurt, 2000, pp. 38–42; M. Nadin, Prologema: What speaks in favor of an inquiry into anticipatory systems? In: R. Rosen, *Anticipatory Systems*, 2nd edition. Springer, 2012 (first published 1985 by Pergamon Press); R. Poli, *Introduction to Anticipation Studies*, Springer, 2017; and others.

[2] R. Rosen, *Anticipatory Systems*, 2nd edition. Springer, 2012, p. 51 (first published 1985 by Pergamon Press).

[3] S. Vrobel, *Fractal Time*. The Institute for Advanced Interdisciplinary Research, Houston, 1998.

[4] For a more detailed philosophical overview about the concept of time from Augustinus via McTaggart, Bieri, Bergson and Husserl to Prigogine, Penrose and Einstein, see, among others, S. Vrobel, *Fractal Time*. The Institute for Advanced Interdisciplinary Research, Houston, 1998; R. Poli, *Introduction to Anticipation Studies*, Springer, 2017.

[5] G. Buzsáki, *Rhythms of the Brain*, Oxford University Press, New York, 2006, p. 6.

[6] M. Nadin, Prologema: What speaks in favor of an inquiry into anticipatory systems? In: R. Rosen, *Anticipatory Systems*, 2nd edition, Springer, 2012 (First published 1985 by Pergamon Press), p. xlvi.

[7] E. Husserl, *Vorlesungen zur Phänomenologie des inneren Zeitbewußtseins.* Niemeyer, 1980, p. 389 (First published in 1928); S. Vrobel, *Fractal Time*. The Institute for Advanced Interdisciplinary Research, Houston, 1998.

[8] H. Lefebvre, *Rhythmanalysis: Space, Time and Everyday Life.* Éditions Syllepse, Paris, 1992 (English translation: Continuum, London, 2004, p. 81).

[9] First English publication in 1998: S. Vrobel, *Fractal Time*. The Institute for Advanced Interdisciplinary Research, Houston, 1998.

[10] B. Mandelbrot, *The Fractal Geometry of Nature*, W. H. Freeman, San Francisco, 1982; L. F. Richardson, The problem of contiguity: An appendix to statistics of deadly quarrels. *General Systems Yearbook.* The Society for General Systems Theory, Vol. 6, Ann Arbor, 1961, pp. 139–187.

[11] S. Vrobel, *Fractal Time*. The Institute for Advanced Interdisciplinary Research, Houston, 1998.

[12] This is meant by the definition of a fractal as a structure whose Hausdorff Dimension exceeds its topological dimension. For a more detailed overview of fractal dimensions, see B.B. Mandelbrot, *The Fractal Geometry of Nature*, W. H. Freeman, San Francisco, 1982; M. Barnsley, *Fractals Everywhere,* Academic Press, London, 1988; Wikipedia: List of fractals by Hausdorff dimension. Accessed 17.07.2010. http://en.wikipedia.org/wiki/List_ of_fractals_by_Hausdorff_dimension; and I. Pilgrim and R. P. Taylor, Fractal analysis of time-series data sets: Methods and challenges. *Fractal Analysis.* By S.-A. Ouadfeul (ed.), IntechOpen 2019.

[13] Reprinted from: I. Pilgrim and R. P. Taylor, Fractal analysis of time-series data sets: methods and challenges. In: S.-A. Ouadfeul (ed.) *Fractal Analysis*. IntechOpen, 2019, CC BY 3.0.

[14] M. Barnsley: *Fractals Everywhere*, Academic Press, London, 1988.

[15] *ibid*, pp. 176–177.

[16] More on the topic of complexity in Chapter 9. See also: S. Vrobel, *Fractal Time. Why a Watched Kettle Never Boils*, World Scientific, 2011; J. L. Casti, *Complexification — Explaining a Paradoxical World through the Science of Surprise*, Harper Perennial, New York, 1995, p. 9.

[17] Reprinted from: I. Pilgrim and R. P. Taylor, Fractal analysis of time-series data sets: Methods and challenges. *Fractal Analysis*. S.-A. Ouadfeul (ed.), IntechOpen, 2019, original in colour, CC BY 3.0.

[18] S. Vrobel, How to make nature blush: On the construction of a fractal temporal interface. In: D. S. Broomhead, E. A. Luchinskaya, P. V. E. McClintock and T. Mullin (eds.) *Stochastics and Chaotic Dynamics in the Lakes: STOCHAOS*, AIP (American Institute of Physics), New York, 2000, pp. 557–561.

[19] B. Mandelbrot, *The Fractal Geometry of Nature*, W.H. Freeman, San Francisco, 1982.

[20] R. F. Voss, 1/f noise and fractals in DNA-base sequences. In: A. J. Crill, R. A. Earnhaw and H. Jones (eds.) *Applications of Fractals and Chaos*, Springer, New York, 1993, pp. 8–9.

[21] https://www.researchgate.net/publication/265152161_Living_in_the_Pink_Intentionality_Wellbeing_and_Complexity/figures. CC BY 4.0.

[22] R. F. Voss, 1/f noise and fractals in DNA-base sequences. In: A. J. Crill, R. A. Earnhaw and H. Jones (eds.) *Applications of Fractals and Chaos*, Springer, New York, 1993, p. 9.

[23] R. F. Voss and J. Clarke, 1/f noise in music and speech. *Nature*, 258, 317–318, 1975; R. F. Voss, 1/f noise and fractals in DNA-base sequences. In: A. J. Crill, R. A. Earnshaw and H. Jones (eds.) *Applications of Fractals and Chaos*, Springer, New York, 1993, p. 10.

[24] G. van Orden, H. Kloos und S. Wallot, Living in the pink: Intentionality, wellbeing, and complexity. In: C. Hooker (ed.) *Philosophy of Complex Systems,* Handbook of the Philosophy of Science, Elsevier, Amsterdam, 2011, pp. 629–672.

[25] L. Nottale, Scale relativity, fractal space-time and morphogenesis of structures. In: H. H. Diebner, T. Druckrey and P. Weibel (eds.) *Sciences of the Interface.* Genista, Tübingen, 2001, pp. 38–51; L. Nottale, Relativité, être et ne pas être. In: *Penser les limites. Ecrits en l'honneur d'André Green*, Delachiaux et Niestlé, Paris, 2002, p. 157.

[26] R. Cash, J. Chaline, L. Nottale and P. Grou, Développement Humain et Loi Log-Périodique. In: *C.R. Biologies* 325. Académie des Sciences/Editions Scientifiques et Médicales Elsevier SAS, 2002, pp. 585–590.

[27] L. Nottale and P. Timar, Relativity of scales: Application to an endo-perspective of temporal structures. In: S. Vrobel, O. E. Rössler and T. Marks-Tarlow (eds.) *Simultaneity –Temporal Structures and Observer Perspectives*, World Scientific, Singapore, 2008, pp. 229–242.

[28] K. Welch, *A Fractal Topology of Time*, Fox Finding Press, 2020 (First published in 2010).

[29] M. S. El Naschie, Young double-slit experiment, Heisenberg uncertainty principle and correlation in Cantorian space-time. In: Mohammed S. el Naschie, O. E. Rössler and I. Prigogine (eds.) *Quantum Mechanics, Diffusion and Chaotic Fractals*, Pergamon, Elsevier Science, 1995, pp. 93–100.

[30] G. N. Ord, Fractal space-time: A geometric analogue of relativistic quantum mechanics. *Journal of Physics A.: Mathematical and General* (The Institute of Physics), 16, 1869–1884, 1983.

[31] M. S. El Naschie, A review of E-infinity theory and the mass spectrum of high energy particle physics. *Chaos, Solitons and Fractals* (Elsevier Science), 19(1), 209–236, 2004; L. Marel-Crnjac, *Cantorian Space-Time Theory*, Lambert Academic Publishing, 2013.

[32] E.g. H. Kantz and T. Schreiber, *Nonlinear Time Series Analysis*, Cambridge University Press, 1997.

[33] S. F. Timashev, Flicker noise spectroscopy and its application: Information hidden in chaotic signals (review). *Russian Journal of Electrochemistry* (MAIK Nautka/Interperiodica, Russia), 42(5), 424–466, 2006.

[34] D. M. Dubois, Incursive and hyperincursive systems, fractal machine and anticipatory logic. In: *CASYS 2002. AIP Conference Proceedings*, Vol. 573, 2001, pp. 437–451.

[35] D. G. Stephen and J. A. Dixon, Multifractal cascade dynamics modulate scaling in synchronization behaviours. *Chaos, Solitons & Fractals* (Elsevier), 44(1–3), 160–168, 2011.

[36] V. Marmelat and D. Delignières, Strong anticipation: Complexity matching in inter-personal coordination, *Experimental Brain Research*, 222, 137–148, 2012.

[37] I. D. Colley and R. T. Dean, Origins of 1/f noise in human music performance from shortrange autocorrelations related to rhythmic structures. *PLoS One*, 14(5), e0216088, 2019.

[38] E.g.: D. G. Stephen and J. A. Dixon, Multifractal cascade dynamics modulate scaling in synchronization behaviours. *Chaos, Solitons & Fractals* (Elsevier), 44(1–3), 160–168, 2011; P. Grigolini, G. Aquino, M. Bologna, M. Lukovic and B. J. West, A theory of 1/f noise in human cognition. *Physica A*, 388, 4192–4204, 2009.

[39] S. Monto, Nested synchrony — A novel cross-scale interaction among neuronal oscillations. *Frontiers in Physiology*, 3, 2, September 2012. Article 384.

[40] *ibid*, p. 5.

[41] Reprinted from: S. Monto, Nested synchrony — A novel cross-scale interaction among neuronal oscillations. *Frontiers in Physiology*, 3, 2, September 2012. Article 384, CC BY 3.0.

[42] *ibid*, p. 1.

[43] A. Washburn, R. W. Kallen, C. A. Coey, K. Shockley and M. J. Richardson, Interpersonal anticipatory synchronization: The facilitating role of short visual-motor feedback delays. In: *Proceedings of the 37th Annual Meeting of the Cognitive Science Society*, 2015, p. 2619.

[44] D. Delignières and V. Marmelat, Strong anticipation and long-range cross-correlation: Application of detrended cross-correlation analysis to human behavioural data. *Physica A*, 394, 47, 47–60, 2014.

[45] A. Washburn, R. W. Kallen, C. A. Coey, K. Shockley and M. J. Richardson, Interpersonal anticipatory synchronization: The facilitating role of short visual-motor feedback delays. In: *Proceedings of the 37th Annual Meeting of the Cognitive Science Society*, 2015, p. 2620.

[46] *ibid*, p. 2622.

[47] *ibid*, p. 2624.

[48] V. Marmelat, D. Delignières, K. Torre, P. J. Beek and A. Daffertshofer, 'Human paced' walking: Followers adopt stride time dynamics of leaders. *Neuroscience Letters* (Elsevier), 564, 67–71, 2014.

[49] B. J. West, *Where Medicine Went Wrong: Rediscovering the Road to Complexity*, World Scientific, Singapore, 2006; J. M. Hausdorff, S. L. Mitchell, R. Firtion, C. K. Peng, M. E. Cudkowicz, J. Y. Wei, A. L. Goldberger, Altered fractal dynamics of gait: Reduced stride-interval correlations with aging and Huntington's disease. *Journal of Applied Physiology*, 82, 262–269, 1997.

[50] The fractal exponents of the stride-time series were estimated by means of Detrended Fluctuation Analysis (DFA). WIKI defines DFA as follows: "In stochastic processes, chaos theory and time series analysis, detrended fluctuation analysis (DFA) is a method for determining the statistical self-affinity of a signal. It is useful for analysing time series that appear to be long-memory processes (diverging correlation time, e.g. power-law decaying autocorrelation function) or 1/f noise." WIKI, https://en.wikipedia.org/wiki/Detrended_fluctuation_ analysis. Accessed 30.3.23.

[51] V. Marmelat, D. Delignières, K. Torre, P. J. Beek and A. Daffertshofer, 'Human paced' walking: Followers adopt stride time dynamics of leaders. *Neuroscience Letters* (Elsevier), 564, 70, 2014.

[52] *ibid*, p. 70.

[53] P. Terrier and O. Dériaz, Non-linear dynamics of human locomotion: Effects of rhythmic auditory cueing on local dynamic stability. *Frontiers in Physiology*, 4, 2012. Article 230; see also: B. J. West, *Where Medicine Went Wrong: Rediscovering the Road to Complexity*, World Scientific, Singapore, 2006.

[54] C. Stetson, X. Cui, P. R. Montague and D. M. Eagleman, Illusory reversal of action and effect. *Journal of Vision*, 5, 769a, 2005; See Chapter 3, Section 3.5.

Chapter 6

Embodied Cognition/Embodied Anticipation

In the 17th century, René Descartes claimed that we manifest our existence in two realms of being: those of *res cogitans* and *res extensa*. The first would form the world of mental activity and the second the physical extension and activity of a person. Descartes' answer to the question of how those two realms might interact led to the famous Cartesian Cut, which was to dominate our scientific paradigms for centuries. It assigned intelligence to *res cogitans* and reduced *res extensa* to an insignificant appendage which does not partake in cognitive acts. Consequently, Cartesian dualism would not have ascribed any anticipative faculties to the body but exclusively to *res cogitans*.

Today's scientific community no longer endorses this dualist worldview. The notion of embodied cognition, which is based on the assumption that our bodily actions and perceptions determine our thinking and behaviour, is now widely accepted. Opinions are divided, however, as to what degree our bodies are involved in our decision-making processes and behaviour and whether we can do without representations. Nevertheless, few would argue that being embedded in a physical, biological and social context is a necessary condition for embodiment. The prevailing definition of embodiment comprises three conditions: The agent is to be situated, extended and distributed [1].

This chapter gives a brief introduction to notions of embodied cognition and embodied anticipation. The assignment of representations to weak anticipation and architectures without representations to strong anticipation (and direct perception) is relativized. The concept of strong embodiment allows for representations, as long as the body is given a clear explanatory role.

Analogously, I shall define embodied anticipation as compensatory action which is explained by bodily causes but allows for representations, including those of the body. This makes embodied anticipation an example of local strong anticipation as witnessed, for instance, in the transformation of extrapersonal into peripersonal space.

Making the act of representation transparent to the obserpant is in itself a strong anticipatory act, as it generates immediacy (see Section 4.2). The world is experienced more directly if the obserpant is capable of compensating the delaying process of recognizing a representation for what it is. The body image generated by Metzinger's phenomenal self-model is classified as an example of strong embodiment, as it is partly based on external and partly on proprioceptive input.

The last part of this chapter revisits the topic of anticipation in artificial agents and assesses how far embedded and situated robots are strong anticipative systems.

6.1 Situated, Extended and Distributed: The Hallmarks of Embodied Cognition

To begin with, I shall briefly outline the three characteristics of embodied cognition. Extended obserpants are situated in an environment with which they interact and which is involved in the obserpant's cognitive functions.

6.1.1 *Situatedness*

French phenomenologist Merleau-Ponty was aware of the body's role in mental activity. He realized that our very situatedness makes our body the blind spot of both perception and cognition within what he called the primordial field of presence. When we see an object such as a house, we see it only from the side which our positive field of perception reveals. However, we assume that there is more to the house than its façade. Because our bodies occupy a certain space for a certain period of time, the world co-exists coupled with our situation. When we perceive the world, we are not aware of our own spatial and temporal perspectives, which forms a phenomenal blind spot [2].

6.1.2 *Extension and distribution*

The view that cognition is extended and distributed was developed, among others, by Noë *et al.* [3] and has since found applications in fields as diverse as robotics and psychotherapy [4]. It basically states that the mind is extended and consists not only of the brain and the rest of the body but can also incorporate parts of the outside world. In addition to assimilated tools and prostheses, internalized external temporal structures, like circadian rhythms or delays to which we have been conditioned, also form obserpant extensions.

The idea that cognition is extended and distributed was implicit in Chapter 3 and exemplified, for instance, in the macaque monkey's control loop. The control loop acted as an extended and distributed system, in which the components monkey, joystick, monitor and robot arm formed a systemic whole. While the delay caused by the extension (the robot arm) was transparent to the macaque, it retained a sense of agency after recalibration. Another example of situated, extended and distributed cognition is the transition from extrapersonal to peripersonal space. A stick for reaching functions as an extension of an obserpant and is part of a systemic whole whose distributed components have merged into one. The obserpant experiences a sense of agency while the accompanying delay and distance compensation remains transparent. In both cases, cognition is extended beyond the brain and the body as it also incorporates part of the environment.

Another extended and distributed phenomenon generated by a situated agent is social trust, which is presented as an example of strong embodiment in Chapter 8.

6.2 Weak and Strong Embodiment

Early models of extended cognition, including the concept of an extended mind, see the brain, the body and the environment as equal contributors to the construction of a coherent world [5]. However, this parity principle, which implies that incorporated nonneural components are also parts of cognitive processes, turns out to be of a rather general nature. As philosophers and psychologists Adrian Alsmith and Frédérique de Vignemont have pointed out, this principle prevents a distinction between cognitive processes which may be outsourced and those which necessarily have to be embodied. Alsmith and de Vignemont argue for a complementarity principle to distinguish between necessarily embodied processes and those which may be attributed to the environment:

> In complementarity views, non-neural elements find their place in a cognitive system in virtue of both being coordinated with neural processes and providing a *different kind* of functionality from the neural processes involved. This is still broadly compatible with conceiving of the mind in terms of cognitive functions, but it puts pressure on the idea that those functions can be realized by just any old (neural or environmental) material. [6]

The researchers proceeded to question whether the existence of body representations and the explanatory role of the body can be reconciled and suggested a differentiation between weak and strong embodiment:

> … we will call any view that gives a clear explanatory role to the body a 'strongly embodied' view (or 'strong embodiment'); by contrast, we will call any view that gives a clear explanatory role to representations of the body, whilst not also giving a clear explanatory role to the body itself, a 'weakly embodied' view ('weak embodiment'). [7]

Why is this important for the concept of strong anticipation? Alsmith and de Vignemont emphasized that strong embodiment allows us to include representations in obserpant extensions, as long as the body is given a clear explanatory role. In the same vein, we can now define embodied anticipation as compensatory action which is explained by bodily causes but allows for representations, including representations of the body. As such, embodied anticipation is an example of local strong anticipation in extended obserpants. Examples of such local strong embodied anticipation are the anticipative adaptive functions of peripersonal spaces.

Strong anticipative behaviour, such as transforming extrapersonal space into peripersonal space, remains elusive if one cannot fall back upon representational models — at least in addition to the body's explanatory role.

Behaviour which cannot be explained without a representational model includes extreme states in which the mind decouples itself partly or completely from its environment and processes internalized structures only, such as being locked in an isolation tank or going through out-of-body experiences. Metzinger's phenomenal self-model is also an example of behaviour which cannot be described without the assumption of a representational model. The body image of the phenomenal self-model is based on sensory, but also on proprioceptive input, a fact which Metzinger exemplifies with phantom pain and neglect (see Chapter 3) as purely internally generated neural activation patterns [8].

Metzinger differentiates between 1st-, 2nd- and 3rd-order embodiment: While the 1st-order embodiment is simply reactive and adaptive, the 2nd-order type is capable of true learning. Only 3rd-order embodiment is able to raise representational content onto a conscious level so that an organism or a robot experiences itself as embodied.

For our purposes, the crux of Metzinger's phenomenal self-model is the fact that this model is transparent to us — we do not recognize it as a model but see through it and experience ourselves and our environment in a naïve realistic perspective as immediate, unfiltered. He stresses that we experience our body as real because the bodily component of our phenomenal self-model is transparent:

> 'Transparency' is here used to refer to an exclusive property of phenomenal states, meaning that the fact of their being representational states is inaccessible to introspective attention. Metaphorically speaking, the construction process is 'invisible' on the level of phenomenal experience itself. Phenomenal transparency explains why we *identify* with the content given via [3rd-order embodiment]. [9]

This transparency of the phenomenal self-model, which Metzinger sees as a condition for the emergence of a first-person perspective, is a shortcut, a compensation, as it bridges an otherwise cumbersome and time-consuming access to the world [10].

As I briefly outlined in Section 4.2, without transparency of the representational model, we would have to first recognize a percept as a representation and then construct a coherent world. Naïve realism spares us this diversion, as it bridges the unnecessarily delaying process of realizing that we are dealing with a representation, not the real thing. Compensating the delaying process of recognizing a representation as what it is generates immediacy and is thus another type of strong anticipation. Against the background of Alsmith and de Vignemont's definition, the body image produced by Metzinger's phenomenal self-model is an example of strong embodiment.

6.3 Observant Extensions Revisited

Embodied strong anticipation includes the anticipative adaptive function of peripersonal space. As briefly described in Chapter 3, artificial extensions to the human body can result in a remapping of extrapersonal to peripersonal space (see Figs. 6.1 and 6.2). And although the use of cross-modal congruency tasks to show that extensions to the body modify peripersonal space has been questioned [11], there is overwhelming evidence for the remapping of extrapersonal to peripersonal space [12].

An important proviso, however, needs to be addressed: Obserpant extensions alone, be they of a physical or virtual nature, do not suffice to cause remapping. Two additional conditions need to be fulfilled: A sense of agency and synchronous coordination. This section looks at a range of remapping examples based on various modes of perception.

6.3.1 *A sense of agency and nested simultaneity*

Psychologist Mariano D'Angelo *et al.* showed that a sense of agency and synchronous coordination are necessary conditions for extending and contracting peripersonal space [13]. In an experiment, they manipulated the sense of agency in participants by means of a virtual hand which

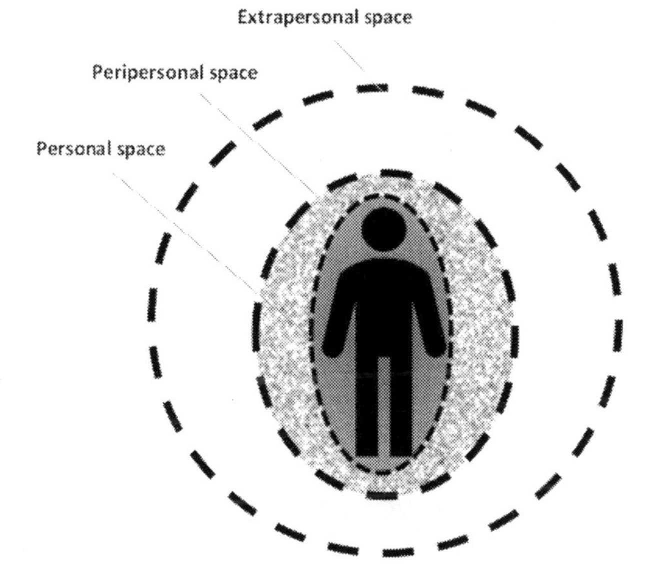

Fig. 6.1. Extrapersonal space is not within reach, but peripersonal space is in reach.

Fig. 6.2. Peripersonal and extrapersonal space are in reach.

extended or reduced their peripersonal space. Afterwards, participants were asked to locate the midpoint of their forearm (which was hidden from view). As expected, their peripersonal space was extended if the virtual hand had been in far space and contracted if it had been closer than the real hand. Note that it was not a sense of ownership but a sense of agency which caused the extension or reduction. This was achieved through active reaching movements, which provided sensory feedback.

Such visuo-tactile interaction is the most common method of determining peripersonal space. This is simply a result of the fact that, although visual and tactile impressions are processed within the body, visual information is assigned to the space outside the body and tactile information is generated within. Binding these two enables us to create a representation of the intermediate space [14].

This intermediate space refers to the space between personal space and the accessible space around it [15]. "Accessible" does not only refer to the intermediate space created through visual and tactile binding. Auditory and tactile binding, as well as social interaction and individual fears, generate such intermediate spaces.

6.3.2 *Remapping into the future*

Cognitive scientist Anna Belardinelli *et al.* conducted experiments in which anticipatory spatial remapping preceded goal-directed action [16]. They questioned whether the peripersonal hand space would be mapped into the future, namely to the position of the anticipated event, such as an object grasp. The cross-modal approach was based on visual and tactile interaction involving visual distractors at the spatial future position of the subjects' finger and correlated gaze behaviour. The results showed that

> ...we map sensory stimulations onto each other via spatial predictive encodings in anticipation of the currently aimed-at next event boundary. (...) Our eyes explore this boundary to ensure that the application of a proper grasp, while the hand begins to expect the upcoming touch of the object before it takes place — and this happens partially even before starting to move. In other words, our results show that [peripersonal space] is remapped into the future for invoking and controlling

> goal-directed manual motor behaviour with respect to a currently intended, future body state. [17]

D'Angelo *et al.*'s conditions for mapping not-yet-reachable space onto reachable space were both met in the experimental setups: a sense of agency and synchronous perception. The peripersonal hand space was created through the simultaneous visual and tactile perceptions and was mapped onto the future position of the hand (when it was going to grab an object). This obserpant extension, which included the intermediate space created through cross-modal congruency, displayed strong anticipative behaviour: A boundary shift towards the outside occurred, the boundary shift resulted in the formation of a new systemic whole, the interfaces which had merged or had newly emerged were transparent to the obserpant and the obserpant retained a sense of agency after the boundary shift and spatio-temporal compensation.

In a different approach, cognitive scientist Tina Iachinia *et al.* showed that spatio-temporal information is processed differently in peripersonal and extrapersonal space. They asked subjects to predict collisions between balls, both inside and outside their peripersonal space. This was done in an immersive virtual reality environment. The participants predicted collisions in peripersonal space more easily than those in extrapersonal space, confirming the assumption that peripersonal spaces play the role of an anticipatory buffer to allow a person or organism to prepare adequate responses in sufficient time [18]. Neuroscientist Giuseppe di Pellegrino and neuropsychologist Elisabetta Ládavas showed that peripersonal space is also represented on the neural level in the fronto-parietal areas, where it displays a high level of multi-sensory integration [19].

6.3.3 *Extended safety margins*

The boundary of peripersonal space can also determine a safety margin around an obserpant. "Looming" stimuli are threatening aspects of the environment and as such provide useful optical cues for hazardous encounters, such as imminent collisions. This was already shown by William Schiff *et al.* in 1962 in experiments with monkeys [20]. Cognitive psychologist Eleonora Vagnoni *et al.* confronted subjects with threatening and nonthreatening images which expanded in size for a second before disappearing. The participants were asked to imagine that the visual stimuli continued to approach them after the image disappeared and judge the point of collision, at which moment they were instructed to press a button. It turned out that participants underestimated the time-to-collision when the stimuli were of a threatening nature (spiders and snakes) but not when the stimulus presented nonthreatening creatures (butterflies and rabbits) [21].

Individual phobias were assessed with a questionnaire in order to determine their effects on time-to-collision estimation. Vagnoni *et al.* found that subjects with spider or snake phobia underestimate the time-to-collision with those creatures. This showed that estimating the time to collision is steered not only by optical cues, as observed by W. Schiff *et al.*, but also by affective content of looming. Thus, Vagnoni *et al.* suggest that the underestimation of the temporal span before collision is a selection effect, as it would allow the threatened individual to respond faster to the threat. The temporal illusion of underestimating the arrival time compensates what would otherwise, i.e. with a correct estimate, have been a delay in response. This strong anticipative behaviour is another example of obserpant extensions which result from delay compensation: From the obserpant's perspective, the length of time, i.e. the temporal dimension measured as succession, contracts (the time-to-collision is underestimated, therefore there is less extension in Δt_{length}). Concurrently, Δt_{depth}, the depth of time, which manifests itself as the temporal dimension measured as nested simultaneity, expands. (Compensating the delay between a correct estimate and the underestimated time makes the collision more immediate and thus increases nested simultaneity between the obserpant and environment.)

Whenever peripersonal space is expanded and compensated, Δt_{length} (succession) is transformed into Δt_{depth} (nested simultaneity) — a manifestation of local strong anticipation in extended obserpants.

The so-called "defensive peripersonal space" is an example of a safety margin which may be used in the assessment of risk in professions which are exposed to looming threat, such as police, firefighting or the military. For instance, the extension of the defensive peripersonal space around the face determines how fast reflexes such as blinking are triggered by an approaching object [22].

A different kind of safety margin is the peripersonal space around cynophobic obserpants (people and other organisms that are afraid of dogs). Unlike spaces of interaction with the external world that are created through visual and tactile cross-modal congruency, their safety margin results from audio-tactile integration [23]. Similar to the experiment involving images of spiders, etc. discussed above, cognitive neuroscientist Marine Taffou and cognitive psychologist Isabelle Viaud-Delmon confirmed that the size of peripersonal space is influenced by anxiety. They asked subjects to react to a tactile stimulus while they were approached by a threatening sound in the form of a dog growling and a nonthreatening sound in the form of a bleating sheep. It turned out that cynophobic participants' peripersonal space was extended in the dog-growling scenario, while there was no significant difference in the extension of the peripersonal space of those who were not afraid of dogs. (The bleating sheep made no impact in either group.) The researchers conclude that anxiety modulates the extension of the peripersonal space, which is in line with Vagnoni *et al.*'s visuo-tactile example [24]. Taffou and Viaud-Delmon suggest that the extension of the peripersonal space may be used to assess the severity of a phobia.

In view of the broad range of descriptions of peripersonal space, de Vignemont and Iannetti propose a dual model of peripersonal space, which differentiates between bodily protection and goal-directed action: Protective space is an interface to avoid threats and thus bodily harm. Working space, on the other hand, is an interface for the body to act on environmental objects [25]. For our purposes, however, either function has the effect of creating an intermediate space which the obserpant compensates. The active and immediate compensation of distance and delay meets my four conditions for local strong anticipation in extended obserpants. The creation of a peripersonal space both through bodily protection and goal-directed action is accompanied by the transformation of Δt_{length} into Δt_{depth}.

6.4 Embodied Anticipation in Robotic Devices

In Chapter 3, where we looked at exoskeletons as components of extended obserpants, Serena Ivaldi, Research Scientist at *AnDy*, pointed out that

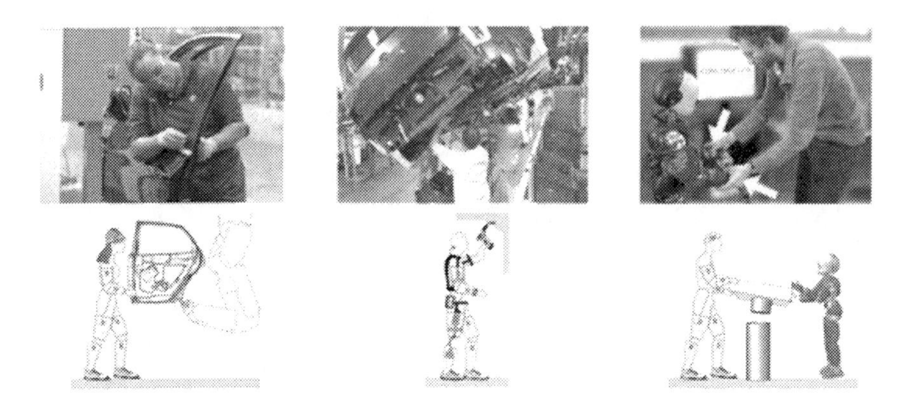

Fig. 6.3. The three collaboration scenarios with the three different robot types: cobot or robotic manipulator, exoskeleton and humanoid robot [27].

anticipation is a necessary condition for successful collaboration between humans and cobots. Cobots (robots which co-exist with humans and share workspaces, see Fig. 6.3) were mainly programmed to avoid collisions and control physical contact with humans. Advanced forms of cobots adapt to human needs and display more advanced skills, such as social interaction demands [26].

Ivaldi describes the complexity of simple human collaboration, which is still difficult for robots to attain:

> If we observe two human beings collaborating, we quickly realise that their synchronous movements, almost like a dance, are the outcome of a complex mechanism that combines perfect motor control, modelling and prediction of the human partner and anticipation of our collaborator's actions and reactions. While this fluent exchange is straightforward for us humans, with our ability to 'read' our human partners, it is extremely challenging for robots. [28]

She exemplifies this task with two humans moving a heavy couch together and stresses that, apart from haptic interaction, they also make use of visual and verbal cues. Collaboration

> … is a complex bidirectional process that efficiently works if both humans have a good idea of the model of their partner and are able to predict his/her intentions, future movements and efforts. [29]

AnDy used wearable sensors and a probabilistic skill model for a collaborative robot. To model human multi-modal communication during collaboration, Ivaldi *et al.* included anticipatory directed gaze to indicate the location of goal-directed actions and haptic clues to start and correct movement. *AnDy* uses *Xsens MVN Analyze* for retrieving postural information and inertial motion tracking to be processed by their machine-learning tools. For this task, the researchers used a pool of data on common movements from which the most likely continuation is deduced [30]. This allows cobots to pick out relevant features to model and predict movement:

> We look at the most probable trajectory that the human is supposed to perform. Imagine you are picking up a heavy box (…) we have the classical recommendation that you use your knee instead of bending your back. But if we see that the human is going to pick up the box while bending the back, we can compute the future trajectory. Let's say in 400 milliseconds, the person is going to complete a bad posture — we can say at that moment of prediction: stop. [31]

Cobots and exoskeletons which can predict movements are better at planning assistance needed by a human collaborator or wearer. Both work on an interactive model which partly draws on external data to estimate future movements.

A cobot which helps a human being to carry a heavy couch up the stairs may form a new systemic whole together with the human and couch. The human and cobot are coupled in their movements and are both active and reactive at the haptic and visual levels. The cobot's anticipatory gaze towards the target location is a goal-directed action but does not implicitly relate to the future — the target location is a set value. The interfacial cut is shifted outwards, as both the human carrier and the cobot "feel" both couch and cobot/human challenged by their environment (e.g. a staircase). However, assigning a sense of agency to the cobot is certainly premature.

So-called ethical robots are based on a "consequence engine" [32] and run on simulation-based models of themselves and their environment. Compared to purely reactive architectures, the consequence engine anticipates not only its own future actions but also those of other robots or human beings it interacts with. It contains a simulator which runs both a robot model and a world model.

Roboticists Alan Winfield and Verena Vanessa Hafner point out the importance of two predictive internal models to establish anticipatory faculties in robotic devices: predictive internal models for sensorimotor control and predictive internal models for safety and ethical systems [33]. If the robotic device anticipates all possible dangers to itself and other robots or humans in its environment, it can be programmed to execute an action which is unsafe for itself but which will save a human being from being harmed. Winfield and Hafner describe an experiment in which they implemented a consequence engine in a robotic device placed in front of a hole in the ground. A human being who was moving towards the dangerous hole was located within the robot's sphere of influence. Then the researchers asked the robot to move either forward, to the right, to the left or to stand still. Only by moving to the right could the robot intercept the human walking towards the hole, which it did, although it ran the risk of damaging itself. Winfield and Hafner conclude that the robot anticipated the consequences for itself and the human being and then applied the ethical rule that humans must be protected at any cost. Robots capable of anticipating human needs will be most valuable as care robots in nursing environments or as cobots at the workplace.

However, simulation-based internal models with such anticipative faculties come at a cost, namely computational overhead. They are not directly coupled with their environment as in enactivist architectures but require representation in the form of an internal model.

One autonomous robotic device which learns and establishes an internal model to anticipate its own motor actions, including their consequences, is the humanoid robot Nao (see Fig. 6.4) [34]. Numerous versions of Nao have been released with specialized programs for educational and nursing environments, tourism and general business tasks. Nao is a bipedal, soccer-playing humanoid robot and household companion. It has been successfully employed in teaching at schools and universities and in working with autistic children. Nao also works at airports and in retail, welcoming visitors and taking over repetitive tasks.

To achieve this, it combines an inverse model which simulates motor commands with a forward model which anticipates its outcome.

Fig. 6.4. Nao evolution [35].

Comparing the envisaged sensory state with the simulated outcome reveals the prediction error (see Chapter 2 on predictive coding).

A simple task for an early version of Nao was set by Schillaci *et al.* [36], who suggested using those two internal models in a decision-making process, where a robot had to decide whether or not a tool for reaching needed to be employed. Robot Nao simulated the reaching action with and without a tool, compared the two and selected the one with the smaller prediction error. Peri-robot space was reachable without a tool and extra-robot space only with the help of one [37].

Yang Ye *et al.* showed that human–robot collaboration profits from the incorporation of Large Language Models, such as *ChatGPT*, in robots. It generates trust in human collaborators and facilitates communication and general interaction with robots [38]. The first humanoid robot to incorporate the Chatbot GPT into his brain is Pepper, a robot designed to be employed in education and a relative of Nao. Pepper was first developed by *Aldebaran* (see Fig. 6.5) and enhanced with *ChatGPT* in cooperation with *Proven Robotics* [39].

Fig. 6.5. Pepper. [41]

In Chapter 2, we attributed strong anticipatory behaviour to the outfielder who catches a ball without having to first calculate the trajectory. Diogo Carneiro *et al.* developed *RALS*, the Robot Anticipation Learning System, which is the first anticipative ball-catching robot control system. It cannot match a human catcher but does not entirely rely on calculating the trajectory from in-flight motion [40]. The system anticipates the state of the ball in terms of position and velocity based on the motions the thrower performs before he or she actually releases the ball. This means that the point of interception can be estimated earlier and the robot can also start advancing in the right direction earlier than it would have been able to if it received information only from the motion of the flying ball.

The anticipation learning system divides the task into two successive phases: the throwing phase (the thrower's hand motions before the ball is released) and the ballistic phase (the ball's in-flight motion). And although the point of interception is calculated in the ballistic phase, whereupon the robot sets itself in motion, the throwing phase helps the robot to save time in the initial waiting phase, when the visual information is being processed:

> During [the ballistic] phase, the trajectory of the free-flying ball is modeled as a parabola, and the interception point is predicted from the estimated curve, which triggers a relevant robot movement. However, to obtain a reliable estimation of the ball trajectory, the robot needs to wait until sufficient visual information is acquired. This may result in insufficient time to move to the catching point, particularly in short-distance scenarios … [42]

By taking account of the thrower's hand motions before he or she releases the ball, the researchers managed to compensate some of the delays inherent in the robot's actions. Unlike the human outfielder, however, it still relies on the time-and-energy-consuming approach of calculating the ball's trajectory.

Inherent delays in robotic devices can be compensated by means of anticipating synchronization. Machine learning engineer Henry Eberle *et al.* stabilized a delay system by coupling a plant and an internal model via slave self-feedback [43]. Apart from the coupling factor, the slave and the master are subjected to the same dynamics. The coupling factor causes the slave to develop faster than the master until the coupling factor approaches 0 and the slave synchronizes with the master's future value [44]. The researchers compared parallel and serial arrangements of the robotic arm plant and the model:

> The 'parallel' system couples the plant itself to an internal model that encodes 'normal' response of the control loop without delays. With both the model and the plant tracking the same target, the plant synchronizes with the future state of the model, anticipating by a sufficient amount to counteract the delays in the real system. The 'serial' system treats the moving target as the master, with an internal model predicting its motion. The control signal that corresponds to this predicted target is calculated and used to control the plant. The output of the plant, subject to the real system delays, forms the 'slave' part of the sensory coupling and ensures the degree of anticipation is always appropriate. [45]

The 'parallel' system proved to be less accurate than the 'serial' one, but it was more robust towards unpredictable delay, thus making it more suitable for tasks in unstructured environments (which resemble the world in which humans move and interact).

The researchers conclude that both arrangements manifest hallmarks of strong anticipation, as they require delays in the underlying control loop in order to induce anticipative behaviour and that anticipation is proportional to the delay.

In Chapter 4, I described Auriel Washburn *et al.*'s research into the human ability to anticipate chaotic, i.e. unpredictable, behaviour in other human beings. In a new approach, Washburn *et al.* investigated whether this ability could also be developed in artificial agents, so as to provide them with the ability to anticipate chaotic behaviour in human beings, again based on anticipatory synchronization induced by feedback delay [46].

The researchers inserted delays into the motion control of an artificial agent, which resulted in a greatly improved ability to predict chaotic human behaviour. Furthermore, the incorporated delay also enabled the artificial agent to synchronize in a way which was reminiscent of natural anticipatory synchronization, such as has been described by Washburn *et al.* in their earlier experiment and in other studies [47].

Anticipating human behaviour presupposes the ability to correctly read body language. What mirror neurons manage in humans, simulation programs attempt in robots: mimicking the environment as a method to train empathy. Media artist Bill Seaman developed, in cooperation with Otto Rössler, the notion of situated "neosentient" intelligent machines. These would display not only anticipatory but also empathic skills:

> ... we consider a *Neosentient* robotic entity to be a system that could exhibit the following functionalities: it learns; it intelligently navigates; it interacts via natural language; it generates simulations of behaviour (it 'thinks' about potential behaviour) before acting in physical space; it is creative in some manner; it comes to have a deep situated knowledge of context through multi-modal sensing; it displays mirror competence. [48]

Such a neosentient machine would recognize itself in a mirror and thus have some concept of a self. It could serve as an endo-obserpant in measurements and, as it would have introspective capabilities, it would form one systemic whole with the measured event. Unlike an exo-observer, it would not be decoupled from the system it observes.

The examples of strong embodiment in humans and robotic devices described in this chapter are mainly model-based approaches, including representations of both body and environment. The nonmodel-based examples of coupled systems and anticipating synchronization by Eberle *et al.* and Washburn *et al.* are examples of local strong anticipation as defined by Stepp and Turvey (see Chapter 2). Strong embodiment and also strong anticipation in extended obserpants also manifest themselves in the mapping of extrapersonal onto peripersonal space. It emerges when extended obserpants first created an intermediate space through a sense of agency and synchronous coordination, which they then compensated by recalibrating their boundaries (delay and distance compensation).

The following chapter describes the creation and development of such intermediate spaces in human beings. They come in the shape of Winnicott's transitional objects and potential spaces and will be described as precursors to strong anticipation.

References

[1] W. Tschacher, Wie Embodiment Thema wurde. In: M. Storch *et al.*, *Embodiment — Die Wechselwirkung von Körper und Psyche verstehen und nutzen*, Verlag Hans Huber, 2006; A. Clark, *Supersizing the Mind: Embodiment, Action and Cognitive Extension*, Oxford University Press, New York, 2008.

[2] M. Merleau-Ponty, *Phenomenology of Perception*, Routledge, New York, 2012 (1945), p. 95 (more on phenomenal blind spots in Chapter 10).

[3] A. Noë, *Out of Our Heads*, Hill and Wang, New York, 2009; A. Clark, *Supersizing the Mind: Embodiment, Action and Cognitive Extension*, Oxford University Press, New York, 2008; D. J. Chalmers, *The Conscious Mind*, Oxford University Press, Oxford, 1996.

[4] R. Pfeifer and J. Bongard, *How the Body Shapes the Way We Think — A New View of Intelligence*, The MIT Press, 2006; W. Tschacher, Wie Embodiment Thema wurde. In: M. Storch *et al.*, *Embodiment — Die Wechselwirkung von Körper und Psyche verstehen und nutzen*, Verlag Hans

Huber, 2006; M. Storch, Wie Embodiment in der Psychologie erforscht wurde. In: M. Storch *et al.*, *Embodiment — Die Wechselwirkung von Körper und Psyche verstehen und nutzen*, Verlag Hans Huber, 2006.

[5] A. Clark, *Supersizing the Mind: Embodiment, Action and Cognitive Extension*, Oxford University Press, New York, 2008.

[6] A. J. T. Alsmith and F. de Vignemont, Embodying the mind and representing the body, *Review of Philosophy and Psychology*, 3(1), 8, 2012.

[7] *ibid*, p. 3.

[8] T. Metzinger, *Being No One: The Self Model Theory of Subjectivity*, The MIT Press, Cambridge, MA, 2004 (First published by MIT in 2003) and personal communication, 2006.

[9] *ibid*, pp. 163–173.

[10] T. Metzinger, *Being No One: Consciousness, the Phenomenal Self and the First-Person Perspective*. Foerster Lectures on the Immortality of the Soul. The UC Berkeley Graduate Council, 2004.

[11] N. P. Holmes, Does tool use extend peripersonal space? A review and re-analysis. *Experimental Brain Research*, 218(2), 273–282, 2012.

[12] A. Berti and F. Frassinetti, When far becomes near: Remapping of space by tool use. *Journal of Cognitive Neuroscience*, 2(3), 418–419, 2000; M. D'Angelo, G. di Pellegrino, S. Seriani, P. Gallina and F. Frassinetti, The sense of agency shapes body schema and peripersonal space. *Nature.com/ Scientific Reports*, 8, 13847, 2018; C. Brozzoli, T. R. Makin, L. Cardinali, N. P. Holmes, A. Farnè, M. M. Murray and M. T. Wallace, Chapter 23: Peripersonal space: A multisensory interface for body-object interactions. *The Neural Bases of Multisensory Processes*, CRC Press/Taylor & Francis, Boca Raton (FL), 2012; G. Rizzolatti, C. Scandolara, M. Matelli, and M. Gentilucci, Afferent properties of periarcuate neurons in macaque monkeys: I. Somatosensory responses. *Behavioural Brain Research*, 2, 125–146, 1981; G. Rizzolatti, C. Scandolara, M. Matelli, and M. Gentilucci, Afferent properties of periarcuate neurons in macaque monkeys: II. Visual responses. *Behavioural Brain Research*, 2, 147–163, 1981.

[13] M. D'Angelo, G. di Pellegrino, S. Seriani, P. Gallina and F. Frassinetti, The sense of agency shapes body schema and peripersonal space. *Nature.com/ Scientific Reports*, 8, 13847, 2018.

[14] C. Brozzoli, T. R. Makin, L. Cardinali, N. P. Holmes, A. Farnè, M. M. Murray and M. T. Wallace, Chapter 23: Peripersonal space: A multisensory interface for body-object interactions. *The Neural Bases of Multisensory Processes*, CRC Press/Taylor & Francis, Boca Raton (FL), 2012.

[15] G. Rizzolatti, C. Scandolara, M. Matelli, and M. Gentilucci, Afferent properties of periarcuate neurons in macaque monkeys: I. Somatosensory responses. *Behavioural Brain Research*, 2, 125–146, 1981; G. Rizzolatti, C. Scandolara, M. Matelli, and M. Gentilucci, Afferent properties of periarcuate neurons in macaque monkeys: II. Visual responses. *Behavioural Brain Research*, 2, 147–163, 1981.

[16] A. Belardinelli, J. Lohmann, A. Farnè and M. V. Butz, Mental space maps into the future. *Cognition*, 176, 65–73, 2018.

[17] *ibid*, p. 72.

[18] T. Iachinia, F. Ruotolob, M. Vinciguerra and G. Ruggiero, Manipulating time and space: Collision prediction in peripersonal and extrapersonal space. *Cognition*, 166, 107–117, 2017.

[19] G. di Pellegrino and E. Làdavas, Peripersonal space in the brain. *Neuropsychologia*, 70, 126–133, 2015.

[20] W. Schiff, J. A. Caviness and J. J. Gibson, Persistent fear responses in rhesus monkeys to the optical stimulus of 'looming'. *Science*, 136, 982–983, July 1962.

[21] E. Vagnoni, S. F. Lourenco and M. R. Longo, Threat modulates perception of looming visual stimuli. *Current Biology*, 9, 22(19), R826-7, 2012.

[22] C. F. Sambo and G. D. Iannetti, Better safe than sorry? The safety margin surrounding the body is increased by anxiety. *The Journal of Neuroscience*, 33(35), 14225–14230, 2013; D. F. Cooke and M. S. Granziano, Defensive movements evoked by air puff in monkeys. *Journal of Neurophysiology*, 90, 3317–3329, 2003; C. F. Sambo, M. Liang, G. Cruccu and G. D. Iannetti, Defensive peripersonal space: The blink reflex evoked by hand stimulation is increased when the hand is near the face. *Journal of Neurophysiology*, 107, 880–889, 2012.

[23] M. Taffou and I. Viaud-Delmon, Cynophobic fear adaptively extends peripersonal space. *Frontiers in Psychiatry*, 5, 2014. Article 122.

[24] E. Vagnoni, S. F. Lourenco and M. R. Longo, Threat modulates perception of looming visual stimuli. *Current Biology*, 9, 22(19), R826-7, 2012.

[25] F. de Vignemont and G. D. Iannetti, How many peripersonal spaces? *Neuropsychologia*, 70, 327–334, 2015.

[26] S. Ivaldi, Anticipatory models of human movements and dynamics: The roadmap of the AnDy project, 2017. https://hal.science/hal-01539731/document; See also Figure 3.1 in Chapter 3.

[27] Reproduced with permission from Serena Ivaldi and Francesco Nori (edited, original in colour). (Serena Ivaldi: Anticipatory models of human

movements and dynamics: the roadmap of the AnDy project, 2017, p. 7. https://hal.science/hal-01539731/document).

[28] S. Ivaldi, Intelligent human-robot collaboration with prediction and anticipation. *Ercim.news* 2018. https://ercim-news.ercim.eu/en114/special/intelligent-human-robot-collaboration-with-prediction-and-anticipation.

[29] *ibid.*

[30] How AnDy Uses Xsenx to Produce Anticipatory Analysis. Interview with Serena Ivaldi, 2020. https://www.movella.com/resources/cases/anticipate-through-how-andy-uses-xsens-to-produce-anticipatory-analysis.

[31] *ibid.*

[32] C. Blum, Self-organization in networks of mobile sensor nodes, PhD thesis, Mathematisch-Naturwissenschaftliche Fakultät, Humboldt-Universität Berlin, 2015.

[33] A. F. T. Winfield and V. V. Hafner, Anticipation in robotics. *Handbook of Anticipation,* Springer, 2018.

[34] Developed by *Aldebaran Robotics* (later rebranded as *SoftBank Robotics*), its first public version appeared in 2008. https://www.aldebaran.com/en/nao; https://en.wikipedia.org/wiki/Nao_(robot).

[35] Wikimedia commons. Accessed 20 April 2023: https://commons.wikimedia.org/wiki/File:NAO_Evolution_.jpg, CC 4.0, original in colour.

[36] G. Schillaci, V. V. Hafner, and B. Lara, Coupled inverse-forward models for action execution leading to tool-use in a humanoid robot. In: *Proceedings of the 7th ACM/IEEE International Conference on Human-Robot Interaction* (HRI 2012), Boston, USA, pp. 231–232.

[37] *ibid.*

[38] Yang Ye, Hengxu You and Jing Du, Improved trust in human-robot collaboration with ChatGPT (preprint April 2023 on *Researchgate*).

[39] https://provenrobotics.ai/the-latest-from-the-world-of-proven-robotics-pepper-with-chatgpt-partnership-with-a-saudi-company/.

[40] D. Carneiro, F. Silva and P. Georgieva, Robot anticipation learning system for ball catching. *Robotics*, 10, 113, 2021.

[41] https://commons.wikimedia.org/wiki/File:Pepper_the_Robot.jpg, CC BY-SA 4.0.

[42] *ibid*, p. 9.

[43] H. Eberle, S. J. Nasuto and Y. Hayashi, Anticipation from sensation: Using anticipatory synchronization to stabilize a system with inherent sensory delay. *Royal Society Open Science*, 5, 171314, 2018.

[44] *ibid*, p. 2.

[45] *ibid*, p. 3.

[46] A. Washburn, R. W. Kallen, M. Lamb, N. Stepp, K. Shockley and M. J. Richardson, Feedback dealys can enchance anticipatory synchronization in human-machine interaction. *PLoS One*, 14(8), e0221275, 2019.

[47] A. Washburn, R. W. Kallen, C. A. Coey, K. Shockley and M. J. Richardson, Interpersonal anticipatory synchronization: The facilitating role of short visual-motor feedback delays. In: *Proceedings of the 37th Annual Meeting of the Cognitive Science Society*, 2015, p. 2619; F. Ramseyer und W. Tschacher, Synchrony in dyadic psychotherapy sessions. S. Vrobel, O. E. Rössler and T. Marks-Tarlow, *Simultaneity — Temporal Structures and Observer Perspectives*, World Scientific, Singapore, 2008, pp. 329–347.

[48] B. Seaman, Unpacking simultaneity for differing observer perspectives. In: S. Vrobel, O. E. Rössler and T. Marks-Tarlow (eds.) *Simultaneity — Temporal Structures and Observer Perspectives*, World Scientific, Singapore, 2008, pp. 256–257.

Chapter 7

The Transitional Object as a Precursor to Strong Anticipation

Intermediate spaces between the obserpant and environment are preconditions for local strong anticipation. These spaces are first created and then compensated by the obserpant. Clark and others have shown that such compensation occurs in various obserpant extensions [1]. Some forms of strong anticipation are tangible and others intangible. The first include incorporated tools such as a stick for pointing, and the latter less tangible behaviour such as embodied trust (more on embodied trust as strong anticipation in Chapter 8). An example of such an intermediate space is the extension which is created by transforming extrapersonal space into peripersonal space. The resulting local strong anticipation in the extended obserpant occurs through recalibration and compensation.

But before compensation can take place, an extended spatio-temporal gap — an intermediate space — must evolve or exist, which the obserpants can assign either to themselves or to their environment. Prior to any such assignment, however, the intermediate space belongs to neither. This is the realm of transitional objects and phenomena which I present in this section as precursors to strong anticipation.

7.1 Transitional Objects as Obserpant Extensions

Paediatrician and psychoanalyst Donald Winnicott defined an intermediate area of experience, a potential space which hosts transitional objects and phenomena:

> I have introduced the terms 'transitional objects' and 'transitional phenomena' for designation of the intermediate area of experience, between the thumb and the teddy bear, between the oral erotism and the true object-relationship, between primary creative activity and projection of what has already been introjected, between primary unawareness of indebtedness and the acknowledgement of indebtedness ('Say: "ta"'). [2]

Winnicott emphasizes the inherent difficulty in the obserpant's task to continuously re-align with our ever-changing environment. Obserpants are embedded agents capable of assigning parts of themselves to the world or parts of the world to themselves. Such a boundary shift separates inside and outside, self and nonself. This section focuses on the employment of transitional objects to relate inner and outer reality, to bridge the gap between self and nonself.

The transitional object is defined by Winnicott as the first non-I. It is an object which could come, early in life, in the shape of a teddy bear or a cuddly blanket. Transitional objects belong neither to the inside nor to the outside: They are interfaces which partake in both realms [3]. As such, they determine how we link inside and outside and thus generate our very personal realities. They manifest themselves as an extended boundary, an interface between the extended obserpant and his or her environment, between me-extensions and not-me.

Winnicott denotes this interface which separates and joins me-extensions and the not-me as a "potential space". The potential space evolves from a feeling of trust, of confidence gained from experience. It spans the realms of obserpant and his or her environment but does not belong solely to either. But not only transitional objects live in potential space. Winnicott sees both play and culture, such as art and religion, as inhabitants of potential space, as they belong neither solely to inner or outer reality. These he refers to as "techniques" rather than objects and denotes them as "transitional phenomena" [4].

As he was primarily concerned with developmental psychology, Winnicott looked at the way an infant creates external reality:

> … the infant is allowed to claim magical control over external reality, a control which we know is made real by the mother's adapting, but the infant does not yet know this. The 'transitional object' or first possession is an object which the infant has created, although at the same time we say this we actually know it to have been a bit of blanket or a fringe of a shawl. [5]

At first, infants are allowed to believe in magical control and their first concept of an external reality appears with the gradual realization that they are not equipped with such powers.

However, before this happens, they create transitional objects and phenomena, such as teddies and tunes. Transitional objects are the infants' first possession, the first object they created themselves. A teddy bear or a cuddly blanket allows the infant to self-soothe in unfamiliar situations or helps him or her to go to sleep. But adults, too, are equipped with potential spaces which may host cultural phenomena and also have evolved through confidence and trust gained from experience. They enjoy the mingling of inside and outside of self-created and given objects and phenomena. Winnicott finds it noteworthy that we take it for granted that an individual can exercise religion and enjoy art without being called mad [6].

The importance of transitional objects and phenomena shows itself in the constant striving to create an external reality. This attempt is not always successful, as we can see in individuals who find it problematic to define their boundaries. Individuals who suffer from borderline syndrome fear both "losing and fusing". They cannot be too close for fear of losing themselves and merging with another person, which results in a loss of self. A characteristic of their condition is the inability to create transitional objects and thereby a potential space, which allows the safe creation of an external reality. Instead, they manoeuver through everyday life by means of intricate rules for "interpersonal distance regulation" [7], a control parameter which is also present in healthy individuals, albeit not in a debilitating manner.

It should be noted that Winnicott, as a developmental psychologist, used the term "transitional" to actually refer to the transition the infant goes through from one developmental state to another [8]. Recent usage, however, does not always imply a succession of developmental stages but usually refers to a transition between inside and outside, between I and non-I [9]. These notions of transitional objects and phenomena are extended interfaces which encompass both part of the obserpants and their environment [10]. Both the obserpants' internal structure and that of their environment contribute to the formation of potential spaces and transitional objects and phenomena. And these undergo adaptive change resulting from the evolving relationship between internal and external states.

All culture lives in potential space, which takes the shape of an extended interface that separates and connects the obserpants and their environment. As Judith Kestenberg and Joan Weinstein pointed out, play and culture are extended into both inner psychic and external reality [11]. Less joyful is a mingling of inside and outside in old age, when people fall back upon transitional objects such as dolls or pets to help them face unfamiliar and potentially threatening situations. Developmental psychologist and social scientist Insa Fooken *et al.* have pointed out the ambivalent character of transitional objects, which can have positive functions such as being emancipating and identity-forming but can also mediate negative effects and even have destructive potential. The "first things" and the "last things" in life, such as a favourite doll in childhood, personal items taken to a nursing home and inherited objects take on very different roles as transitional objects in transitional contexts [12].

7.2 Compensation of Obsolete Transitional Objects Is Strong Anticipation

According to Winnicott, absence and reappearance form the first components of an external reality. Through play, the infant experiences that momentarily absent objects tend to reappear. If a hidden cuddly blanket or the mother reappears after vanishing, the infant becomes confident and tolerates a temporary absence. Anticipating the loss and reappearance of the lost object creates a first notion of an external reality, in which objects can appear and disappear [13].

As the infant grows up, transitional objects may become obsolete, but they are not lost, as their meaning and function survive as internalized structures [14]. This means that the loss is compensated by the obserpant, who then does not miss the object, as the gap the object used to bridge is now internalized as a structure which manifests itself as self- and object representations [15].

So what happens in terms of interfacial shifts when a transitional object becomes obsolete? Initially, the gap between inside and outside is created as a potential space which hosts a transitional object. Once the transitional object no longer has a function for the obserpant, its interfacial extension is compensated. The compensatory and recalibrating act manifests itself as strong anticipation, as a once-existing gap between inside and outside is now closed. The potential space which hosted the transitional object continues to shrink until it disappears altogether. However, the meaning and function of the obsolete transitional object endure as internalized structures in the extended obserpant. In a human obserpant, such internalized structures manifest themselves as smooth interactions with the environment and almost frictionless, graceful movements which point to a strong anticipatory faculty [16].

Transitional objects and the accompanying potential spaces serve as precursors to strong anticipation: Compensation is achieved by the system as a whole, as the transitional object and the potential space were created by both the obserpant and his or her environment. Transitional objects partook in both, so their compensation also extends beyond the individual components. The internal structure of the systemic whole consists of an environment and an obserpant who is coupled to and coordinates with this environment through transitional objects or phenomena. Neither the obserpant nor his or her environment need to be aware of what they are doing. Both components just act naturally: They "do what they do". Although Winnicott's terminology did not include obserpant extensions which manifest as delay or distance compensation, one may see transitional objects as such [17].

Intangible transitional objects and phenomena such as behaviour patterns, art and religion are also temporal extensions to an obserpant who has compensated the potential space in which these phenomena live. Compensation is achieved through internalizing the structures of transitional objects and phenomena whose temporal extensions exceed the obserpant's Now and which structure the obserpant's body image:

> Transitional objects differ from all others because they contain elements of the past and present and are bridges to the future. (...) The transitional object is an external aid in the integration of various body parts, rhythm, and shapes into a three-dimensional image of the body. [18]

As extended interfaces, transitional objects and phenomena comprise rhythms both inside and outside the body. These temporal extensions are nested, i.e. they manifest themselves in a fractal arrangement: Neural oscillations are embedded in slower, metabolic ones, which again are embedded in circadian and seasonal rhythms, etc. If compensated, i.e. successfully internalized, these transitional phenomena in the form of rhythms have become obserpant extensions. Embedding rhythms creates nested simultaneity and thus Δt_{depth}. Rhythms without contextualization, i.e. without embedding, extend in the direction of Δt_{length} only. Within an obserpant's Now, or any other inertial system, Δt_{length} is transformed into Δt_{depth} whenever external rhythms are embedded (rather than arranged as successive percepts). Thus, internalizing obsolete transitional objects transforms Δt_{length} into Δt_{depth}.

Other human beings can also function as transitional objects, which are incorporated into an extended obserpant. Physician Volkmar Glaser describes so-called "transcending boundaries", which arise when we incorporate certain parts or aspects of other human beings or fuse with an object. This is the case if, for instance, the hand of a physician or healer is felt as being part of one's own body [19]. The extension is accompanied by an interfacial shift outward. Glaser showed that breathing, circulation and elasticity of tissues improve when we imagine that another person's body or parts of it belong to our own body. Deep and smooth breathing is accompanied by an attitude of trust, which itself is an obserpant extension (see Chapter 8). Glaser called this ability to extend our proprioception to objects and other living beings "transsensus". Incorporated objects include the skis of a skier, for instance, who will feel the consistency of the snow and the obstacles on the ground through his or her skis. As bodily extensions, they will become more and more transparent to the skier and thus form a new systemic whole consisting of skis and skier. Transsensus manifests itself as a compensated obserpant extension which includes a sense of agency and synchronous (i.e. simultaneous) coordination and is thus an example of a strong embodiment. In terms of fractal time, Δt_{depth}, i.e. nested simultaneity, increases with transsensus as an obserpant extension, while Δt_{length} decreases [20].

The first contextualization occurs in the creation of the transitional object, as it belongs to both the internal and the external world. With every new contextualization, every new obserpant-extending temporal embedding, simultaneity expands our Now. The Now or any other inertial system with an endo-perspective contains components of both past and present. For instance, the tear stains on a cuddly blanket will remind an infant of a self-soothing process whose memory creates a bridge to the future, nesting retensions and protensions in a cascade of temporal depth.

Barkin *et al.* remark that reality generation through the creation and compensation of transitional objects is a continuous process which lasts a lifetime:

> …reality is no longer seen as a steady backdrop but rather as a dynamic oscillation of figure and ground; and the organism as a whole is viewed as an open system continuously engaged in mutual development with the outside. [21]

The compensation of transitional objects is a form of strong anticipation: In the act of compensation, the obserpant and the transitional object form one systemic whole which comprises the potential space. Once the structures and meaning of the object which previously resided in potential space have been internalized, a boundary shift towards the outside has occurred. This interfacial shift is transparent to the obserpant, yet he or she retains a sense of agency after recalibration.

Transitional objects themselves live in potential spaces which evolve through confidence and trust gained from experience. As such, they may be regarded as precursors to strong anticipation [22]. However, trust gained from experience can not only create potential spaces and thus precursors to strong anticipation. The following section looks at embedded trust, which itself ranks as an example of strong anticipation.

References

[1] A. Clark, *Supersizing the Mind. Embodiment, Action and Cognitive Extension*, Oxford University Press, New York, 2008; A. Noë, *Action in Perception*, MIT Press, Cambridge, 2004; S. Vrobel, CASYS '09: The extended mind: Coupling environment and brain. Daniel M. Dubois (ed.) *International Journal of Computing Anticipatory Systems* (CHAOS, Liège, Belgium), 2010.

[2] D. W. Winnicott, *Playing and Reality*, Tavistock Publications, 1971, p. 2.

[3] L. Barkin, The concept of the transitional object. Between reality and fantasy. In: S. A. Grolnick and L. Barkin (eds.) *Between Reality and Fantasy. Transitional Objects and Phenomena*, Jason Aronson Inc., 1978, pp. 513–538.

[4] D. W. Winnicott, The location of the cultural experience. In: P. L. Rudnytsky (ed.) *Transitional Objects and Potential Spaces — Literary Uses of D. W. Winnicott*, Columbia University Press, 1993.

[5] D. W. Winnicott, *Human Nature*, Free Associations Books, London, 1988, pp. 106–107.

[6] *ibid*, p. 107.

[7] R. A. Lewin and C. Schulz, *Losing and Fusing — Borderline Transitional Object and Self Relations*, Jason Aronson, New Jersey, 1992.

[8] D. W. Winnicott, *Human Nature*, Free Associations Books, London, 1988, pp. 106–107; D. W. Winnicott, The location of the cultural experience. In: P. L. Rudnytsky (ed.) *Transitional Objects and Potential Spaces — Literary Uses of D.W. Winnicott*, Columbia University Press, 1993.

[9] E.g.: A. Sugarman and L. A. Jaffe, A developmental line of transitional phenomena. In: M. Fromm and B. Smith (eds.) *The Facilitating Environment: Clinical Applications of Winnicott's Theory*, International Universities Press, Madison, CT, 1989, pp. 88–129.

[10] S. Vrobel, Negotiable boundaries. In: R. Trappl (ed.) *Cybernetics and Systems 2010. Proceedings of the European Meeting on Cybernetics and Systems Research*, Austrian Society for Cybernetics Studies, Vienna, Austria, pp. 224–229.

[11] J. S. Kestenberg and J. Weinstein, Transsensus-outgoingness and Winnicott's intermediate zone. In: S. A. Grolnick and L. Barkin (eds.) *Between Reality and Fantasy. Transitional Objects and Phenomena*, Jason Aronson Inc., 1978.

[12] I. Fooken, A. Depner and U. Pietsch-Linst, Betwixt things: The ambivalence of objects in transitional contexts. *Zeitschrift für Soziologie der Erziehung und Sozialisation*, 149, January 2016.

[13] D. W. Winnicott, *Human Nature*, Free Associations Books, London, 1988, p. 106.

[14] S. Vrobel, Transitional objects and transcended boundaries as observer extensions. Talk given at *InterSymp: 21st International Conference on Systems Research, Informatics and Cybernetics. Conference Proceedings*, Tecumseh, Canada, 2009.

[15] L. Barkin, The concept of the transitional object. Between reality and fantasy. In: S. A. Grolnick and L. Barkin (eds.) *Between Reality and Fantasy. Transitional Objects and Phenomena*, Jason Aronson Inc., 1978, p. 525.

[16] V. Glaser, *Eutonie. Das Verhaltensmuster des menschlichen Wohlbefindens.* Lehr- und Übungsbuch für Psychotonik, 4th edition, Karl F. Haug Fachbuchverlag, 1993.

[17] S. Vrobel, *Winnicott's Transitional Objects and Potential Spaces: Internalization of the Exterior and Externalization of the Interior.* Talk given at University of Ghent, Belgium, April 2010.

[18] J. S. Kestenberg and J. Weinstein, Transsensus-outgoingness and Winnicott's intermediate zone. In: S. A Grolnick and L. Barkin (eds.) *Between Reality and Fantasy. Transitional Objects and Phenomena*, Jason Aronson Inc., 1978, p. 76.

[19] V. Glaser, Eutonie. *Das Verhaltensmuster des menschlichen Wohlbefindens.* Lehr- und Übungsbuch für Psychotonik, 4th edition, Karl F. Haug Fachbuchverlag, 1993.

[20] S. Vrobel, *Fractal Time — Why a Watched Kettle Never Boils*, World Scientific, Singapore, 2011.

[21] L. Barkin, The concept of the transitional object. Between reality and fantasy. In: S. A. Grolnick and L. Barkin (eds.) *Between Reality and Fantasy. Transitional Objects and Phenomena*, Jason Aronson Inc., 1978, p. 549.

[22] S. Vrobel, The transitional object as a precursor to strong anticipation. D. M. Dubois (ed.) *International Journal of Computing Anticipatory Systems* (CHAOS, Liège, Belgium), 2010; The concepts and arguments presented in this section are based on talks given at *InterSymp* in Baden-Baden 2009 and at the *Winnicott Workshop* in Ghent 2010.

Chapter 8

Trust as Weak or Strong Anticipation

Obserpant extensions come in many guises, such as delays or distances which, if compensated, manifest themselves as embodied and strong anticipation. Some imply a modification in our sense of proprioception, as do extensions resulting from Glaser's *transsensus* — the ability to sense beyond oneself by merging with parts of our environment (see Chapter 7). By extending our proprioception beyond the boundary of our epidermis, we increase peripersonal space. Physical extensions, like the relaxing touching hand resulting from *transsensus*, change our muscle tone, which also induces a feeling of trust. But trust also emerges from logical reasoning, from adopting abstract frameworks which are mediated via symbol systems.

In the following section, two types of trust are differentiated. The first, symbolic trust, which results from a reduction of external complexity through symbol systems, manifests itself as weak anticipation. The second, embodied trust, results from a reduction of external complexity through physically compensated delays or distances and manifests itself as strong anticipation. If weak anticipation is embodied and compensated, the interpreting obserpant transforms it into strong anticipation, which eventually evolves into habits.

8.1 Embodied and Symbolic Reduction of Complexity

The external world usually has a higher degree of complexity than the embedded obserpant: It contains more options than the obserpant can anticipate. And as both the obserpant and his or her environment form one systemic whole, the relationship between internal and external complexity is a measure of how well the obserpant anticipates environmental change. But anticipation is a bidirectional phenomenon, i.e. the environment, too, anticipates changes in the embedded obserpant. Together, they form what social scientist and semiotician Gregory Bateson denoted as a *unit of survival*, whose flexibility depends on both internal and external complexity [1].

The range of possible responses within a unit of survival may vary and internal and external complexity do not necessarily match. In fact, there is usually a gap which needs to be bridged by the obserpant in order to navigate the world smoothly. Sociologist Niklas Luhmann observed that if the degree of internal complexity does not match that of external complexity, we reduce external complexity through symbol systems by partly shifting the problem from the outside to the inside [2]. According to Luhmann, this symbolic reduction of complexity manifests itself as (social) trust.

Obserpant extensions include not only spatial and temporal extensions but also cultural conventions, such as linguistic and semiotic systems. While spatio-temporal obserpant extensions reduce complexity through physically compensated delays or distances, language and other conventions result from symbolically compensated differences between internal and external complexity. These two forms of complexity reduction correspond to two types of trust — embodied and symbolic — which both reduce environmental complexity to the obserpant.

Niklas Luhmann's definition of a system focuses on two intertwined processes. On the one hand, there is the interface between agent and environment and, on the other hand, interfaces within the system's internal structure, that is,

> … the differentiation of the system from its environment, i.e. the demarcation of a boundary, and the internal differentiation of the system, i.e. the functional specification of its subsystems and mechanisms. [3]

Systems constitute themselves by differentiating between inside and outside and maintain themselves by stabilizing this boundary. As we have seen, these boundaries can be highly negotiable. The maintenance process manifests itself in boundary shifts and constant recalibration. If the environment has a higher degree of complexity than the embedded system, it contains more options than the system can anticipate. The system compensates this difference in complexity by means of a subjective world design by interpreting the world in a selective manner and generalizing the impressions it registers. It thereby reduces the environmental complexity to a level it can handle. As the difference between the environmental complexity and the complexity of the embedded agent decreases, the formation of trust sets in. If the reduction of complexity is carried out intersubjectively, knowledge is generated which is labelled 'true':

> Trust reduces social complexity by overstretching given information and generalizing anticipated behaviour. It does so by substituting a lack of information through an internally guaranteed certitude. [4]

Luhmann's description of trust as a reduction of complexity is based not only on intellectual, cognitive capacity but also on emotions. He argues that emotions are diffuse and cause both an internal and external reduction of complexity:

> [Emotions] reduce environmental potentialities by way of preference for a specific object and thereby reduce [the obserpant's] internal possible range of options to process experiences. [5]

He rates emotional reduction just as vital as symbolic reduction:

> Any collapse of the emotional relationship would re-install the suffocating complexity of this world. [6]

When we encounter differences in internal and external complexity, we deal with this problem on a symbolic level by partly shifting the problem from the outside to the inside. By contrast, embodied trust shifts outward the interface between self and nonself, as it involves incorporation of parts of the environment.

What does embodied trust do to our perspective? It reduces the complexity of our Now, our fractal temporal interface, by incorporating external spatial structures and temporal rhythms, i.e. by increasing Δt_{depth}. Embodied trust may thus also be defined in terms of nested interfaces. In social systems, we tend to trust fellow members or colleagues more than strangers [7]. Like an onion, we maintain nested layers of trustworthiness, whose inner layers consist of close friends, teammates and family, and the outer ones of acquaintances we do not fully trust. The nested layers may result from a shared range of possible internal responses and, in the wake of it, shared compensated delays and distances.

Embodied trust is an example of local strong anticipation: A boundary shift towards the outside generates a systemic whole consisting of the obserpant and part of the environment. The act of compensating the incorporated delay or distance is transparent to the obserpant. And after recalibration, the obserpant retains a sense of agency.

Symbolic trust also reduces interfacial complexity, but it does so by generalizing and thus simplifying external states and processes by means of symbol systems. As representations which are interpreted by an obserpant, they act as shortcuts between inside and outside. However, symbolic trust never shifts the boundary outward, since the compensatory act which reduces external complexity does not involve any obserpant extensions. Furthermore, the symbol systems from which symbolic trust emerges are context-independent, i.e. they are not linked to a specific temporal and spatial obserpant perspective. As arbitrary conventions, symbol systems generate an atemporal type of trust which corresponds to weak anticipation: an external set of rules an interpretant uses to anticipate and steer future events.

8.2 The Obserpant Turns Time-Independent Symbol Systems into Temporal Ones

Both embodied and symbolic trust result from a reduction of complexity. Simplification emerges from compensated delays and distances (embodied trust) and compensated differences in internal and external complexity by means of symbolization (symbolic trust).

Embodied and symbolic trust formation can be modelled on the basis of my Theory of Fractal Time. Whereas embodied trust accumulates as Δt_{depth}, i.e. as nested simultaneity, symbolic trust is based on successive (Δt_{length}-generating) semiotic strings, which in themselves are time-independent. These strings, however, have to be interpreted by an obserpant by contextualizing the string within his or her Now. And as a result of this contextualization, the obserpant creates more Δt_{depth}. Thus, a previously time-independent string turns into a temporal one, when it becomes part of the internal structure of the obserpant's Now, which has turned succession into simultaneity.

A brief excursion into biosemiotics may prove helpful to illustrate the interaction between embodied and symbolic extensions. Developmental Systems Theory states that

> … species-specific traits are formed with the help of structured sets of developmental resources (…). Some of the developmental resources are genetic, whilst most others — from the cytoplasmic machinery of fertilized eggs to the social structurings that influence human psychological development — are nongenetic. [8]

Molecular biologist Jesper Hoffmeyer emphasizes that organisms inherit not only their DNA but also a fitting environment in which the organism's genetic material will thrive. The individual organism — the phenotype — interacts with its environment and optimizes it for the next generation [9]. The phenotype extends in spatio-temporal dimensions, whereas the genotype extends into a string of symbols. However, they come together in the acting and experiencing obserpant which processes the nontemporal string of symbols within a temporal framework.

Developmental Systems Theory calls into question the dichotomy between the organism and its environment, a position that dovetails with Bateson's *unit of survival* and the enactive paradigm. Analogous to making both the extended obserpants — including their compensated delays and distances — and their environment the object of study, we must treat both the symbol systems with which the obserpants reduce their external complexity and their environment as one systemic whole.

8.3 Embodied and Symbolic Trust: Analogue vs Digital Codes

But although we may treat them as one systemic whole, no boundary shift in the form of an obserpant extension correlates with symbolic trust. The shift is inward, as it constrains the obserpant's range of possible responses (which would have been triggered more extensively by a not-yet-simplified environment).

Embodied and symbolic trust are encoded differently. Embodied trust is of a spatio-temporal nature while symbolic trust is based on language and other semiotic and social conventions. Hoffmeyer's definition of two types of coding provides a fitting differentiation between embodied and symbolic trust:

> I will use analog coding as a common designation for codings based on some kind of similarity in the spatio-temporal continuity, or on internal relations such as part-to-whole, cause-and-effect. Digital coding, in contrast, will be used to designate sign systems where the relations of sign to signified are due to a demarcation principle of purely conventional or habitual origin. [10]

Hoffmeyer stresses the importance of an approach which takes account of code duality. A code may be analogue in one context and digital in another. For instance, hieroglyphs can function both as ideograms and phonograms [11].

Analogue coding is iconic and indexical; it implies part–whole relations as well as cause-and-effect and during relations and provides a spatio-temporal continuity. Digital coding, by contrast, is symbolic. It is time-independent and based on arbitrary symbol systems, which are themselves based on convention or habit. Both analogue and digital coding reduce the degree of complexity between obserpant and environment: Complexity reduction via analogue coding results from compensated delays or distances, whereas digital coding reduces complexity through symbol systems.

Using Hoffmeyer's categorization, embodied trust is based on analogue, symbolic trust in digital coding. While embodied trust shares the

Table 8.1. Similarities and differences between the embodies and the symbolic models of trust [13].

Types of trust	Embodied	Symbolic
Mode	Analogue	Digital
Mediation	Direct perception	Representation
Complexity reduction	Complexity reduction through compensated delays/distances	Complexity reduction though symbol systems
Context dependence	Context-dependent	Context-independent
Temporal status	Spatio-temporal continuity, internal part-whole relations or cause-effect and during relations	Time-independent and based on arbitrary sign systems, which are based on convention or habit
Temporal dimension	Δt_{depth} (nested simultaneity)	Potential Δt_{length} (potential succession)
Strong or weak anticipation	Strong anticipation through outward boundary shift	Weak anticipation through inward boundary shift (N.B.: obserpant creates strong anticipation through contextualization)
Sphere of influence	Local	Nonlocal/global

spatio-temporal continuity of the environment and thus creates Δt_{depth}, symbolic trust is temporally stable, i.e. it can remain uncontextualized and therefore generates potential Δt_{length} only. (Table 8.1 lists further differences and similarities between the embodied and the symbolic models of trust.)

We don't have to go as far as the genome as an example of a digital code which is time-conservative (i.e. providing temporal stability and therefore suitable for memory storage), has the potential for combinatorics to renew the code and has the capacity for abstraction (i.e. forming meta-languages). Any mutually developed symbol system, such as human language, will suffice, as it possesses all of these properties.

The complementary analogue code comes in the shape of an obserpant and does not possess temporal autonomy. The obserpant possesses spatial and temporal extensions and is embedded in a spatio-temporal environment. Therefore, he or she is context-dependent, temporally unstable and in a position to contextualize the digital code. While embodied trust creates Δt_{depth}, symbolic trust is temporally stable, i.e. it remains uncontextualized and therefore generates potential Δt_{length} only [12].

Symbolic trust provides a meta-language which will endure in the memory even if it is not recalled and which provides the envirotype for the obserpant. In this state, it is context-independent, temporally stable and nonlocal. As soon as the memory is retrieved by an interpretant, though, he or she shapes the analogue code within his Now. The symbolic message interpreted in the obserpant's Now shapes the internal structure of this Now: Interpretation turns a context-independent and temporally stable code into a context-dependent, local and temporal one. The skills of the interpreting agent will increase with every contextualization and allow him or her to transform symbolic into embodied trust and thus weak into strong anticipation.

Both complexity reduction via compensated delays and distances as well as complexity reduction via symbol systems generate interfacial regularities, i.e. habits. In the latter case, weak anticipation is first transformed into strong anticipation and then evolves into habits via the interpreting embodied obserpant.

References

[1] G. Bateson, *Steps to an Ecology of Mind*, Ballantine, New York 1972.

[2] N. Luhmann, *Vertrauen: ein Mechanismus der Reduktion sozialer Komplexität*, 4th edition. Lucius & Lucius, Stuttgart 2000.

[3] *ibid*, p. 120 (my translation).

[4] *ibid*, p. 126 (my translation).

[5] *ibid*, p. 106 (my translation).

[6] *ibid*, p. 106 (my translation)).

[7] *ibid*, pp. 120/21 (my translation).

[8] J. Hoffmeyer, Biosemiotics: An Examination into the Signs of Life and the Life of Signs. University of Scranton Press, London, 2008, p. 105.

[9] *ibid*, p. 106.

[10] *ibid*, p. 89.

[11] *ibid*, p. 90.
[12] S. Vrobel, Trust as embodied anticipation. Talk given in my symposium *Models of Embodied Cognition*. Operations Research (EURO 2012), Vilnius, Lithuania, July 2011.
[13] S. Vrobel, Fractal Time — Why a Watched Kettle Never Boils, World Scientific, Singapore, 2011.

Chapter 9

Contextualization and Decontextualization

The fractal Now was exemplified in Chapter 5 by Husserl's explanation of why we perceive a tune rather than a succession of uncorrelated notes. It would be inconceivable to hear a tune unless a simultaneous layer provided a context, a framework against whose background successive events may be arranged. Nonfractal obserpants, i.e. organisms without a temporal fractal interface, would indeed perceive only a succession of isolated notes. Because they would lack the nestings which provide the simultaneous contrast necessary to form a temporal fractal perspective, nonfractal obserpants would live in an eternal succession of uncorrelated Nows. This means that neither memory formation nor anticipative behaviour could take place, as both require extensions in Δt_{length} and Δt_{depth}.

Fractal obserpants, on the other hand, consist of nested multiple time scales on which a cascade of internal rhythms, from neural oscillations to cardiovascular and hormone cycles, takes place. In addition, they are embedded in external rhythms, such as the tides or circadian clocks. Therefore, they are able to contextualize and perceive a tune.

However, contextualization can also be experienced as unwanted interference which prevents us from focussing on vital aspects. Simultaneous contrasts are an example of such contextual interference.

But then, reduced Δt_{depth}, i.e. less context, may lead to auditory or visual illusions, leaving the Now deprived of its anticipatory faculty.

The notion of Interface Complexity (IC) is defined to measure the relation between internal and external complexity. For local strong anticipation, IC is measured as the relation between embodied and embedding parameters, and for global strong anticipation, it is measured in the fractal dimension as the relation between embodied and embedding long-term correlations.

9.1 Interface Complexity and the Range of Possible Responses

Mathematician and futurologist John Casti stressed the fact that perspective and context determine the complexity of a structure. Complexity is a subjective concept which only makes sense if applied to systemic wholes, not to isolated systems:

> … the complexity of a political structure, a national economy or an immune system cannot be regarded as simply a property of that system taken in isolation. Rather, whatever complexity such systems have is a joint property of the system *and* its interaction with another system, most often an observer and/or controller. [1]

He suggests that the complexity of a system is proportional to the number of nonequivalent descriptions of that system an observer can generate:

> Suppose our system (…) is a stone on the street. To most of us, this is a pretty simple, almost primitive kind of system because we are capable of interacting with the stone in a very circumscribed number of ways. We can break it, throw it, kick it — and that's a about it. Each of these modes of interaction represents a different (i.e. inequivalent) way to interact with the stone. But if we were geologists, then the number of different kinds of interaction available to us would greatly increase. In that case, we could perform various sorts of chemical analyses on the stone, use carbon-dating techniques on it, x-ray it and so on.

> For the geologist, the stone becomes a much more complex object as a result of these additional — and inequivalent — modes of interaction. We see from this example that the complexity of the stone is a relative matter, dependent on the nature of the system with which the stone is interacting. [2]

If both perspective and context determine the degree of complexity, can we assign complexity to context alone? Astrophysicist Rosolino Buccheri pointed out that understanding complexity as a subjective measure raises the question concerning the status of emergence:

> … it seems clear that any increase in complexity always corresponds to the emergence of new, more organized systems and higher-level rules governing their behaviour. The following question can therefore be raised: is 'emergence' always epistemological or may we have ontological emergences too? [3]

If we place complexity into the eye of the beholder, emergence, too, is perceived through an endo-perspective and thus the result of a private interfacial cut. However, a self-organizing environment which acts as a constraint on the enactive obserpant may cause him or her to suspect an ontological side to emergence. (In Chapter 12, the epistemological and ontological character of fractal spacetime is discussed as a result of co-evolution.)

An example of how the relationship between internal and external constraints can be measured and captured in a single control parameter was devised by developmental psychologists Esther Thelen and Linda Smith. They looked at babies' stepping behaviour and determined the ratio between a baby's leg strength (as the embodied constraint) and leg weight (as the embedding constraint) [4]. The leg weight stands for embedding constraints in the baby's relation to its environment, i.e. relevant environmental dispositions, such as gravity. The leg strength is an embodied constraint of the baby, i.e. its ability to interact with environmental provisions [5]. The relationship changes as the baby develops. At first, stepping is easy because in spite of their lack of strength, the legs are quite light. When the legs become heavier because the baby has gained weight, stepping is more difficult and ceases. It returns later when the baby has gained strength [6].

In the same vein, mathematician Bruce West emphasizes that the complexity of an organism can be expressed in the range of possible responses it has at its disposal. If this range of responses is tuned well so that it is able to handle stress, the organism is healthy [7]. As opposed to traditional views, which have defined disease as a disruption of regularity, he defines disease as a loss of complexity, which manifests in the onset of regularity. A loss in complexity is visible in time series as a loss in variability and can be measured in the fractal dimension [8].

West suggests we shift our attention from allometric indicators to fractal ones, which take account of the fact that human physiology has a nested structure. He presents an abundance of examples of fractal physiology, from the nervous system to cardiovascular, gastrointestinal and gait dynamics. The human body consists of geometrical fractal structures such as can be seen in lung tissue or kidneys, which lead to fractal physiological time series [9].

Statistical fractals in particular are of interest, as they are manifestations of long-term correlations. West concludes that global measures in the shape of statistical fractal time series are better indicators of health or disease than local allometric ones. Therefore, he urges that, rather than measuring the rate of change of heartbeat, breathing rate or stride interval, we should focus on the global measures which measure their variability.

One such indicator of health is heart rate variability, which measures the variability of the time interval from one heartbeat to the next. The time series of a healthy individual has a fractal dimension between 1.1 and 1.3. Higher dimensions indicate uncorrelated processes, lower ones rigid, all-too regular dynamics: Atrial fibrillation correlates with a time series with a fractal dimension of approximately 1.5, and congestive heart failure manifests in a time series with a fractal dimension of 1.0 [10]. Heart rate variability (HRV) and stride rate variability (SRV) are long-term correlations which, for SRV, consist of fluctuations with a long-term memory that spans intervals of over a hundred steps [11]. The SRV of an average walker in a log-log plot (variance/mean) produces a curve with a fractal dimension of 1.30. Higher dimensions would represent uncorrelated random noise, lower ones a regular deterministic process [12].

What is of importance is not the average value of the rate of change in a given parameter but the range of variation in the fractal dimension. Looking at stride range variability (SRV) as an indicator of health, West found that the average fractal dimension of healthy adult individuals' gait lies in the range $1.2 \leq D \leq 1.3$. Children under the age of five, however, showed a wider range of $1.12 \leq D \leq 1.36$. West concludes the following:

> It is clear that the average fractal dimension over each group is the same, approximately 1.24, but the range of variation in the fractal dimension decreases significantly with age. This would seem to make the fractal dimension an increasingly reliable indicator of the health of the motor-control system with advancing age. [13]

If we now look at the embedding environmental constraints which limit the range of embodied responses, it turns out that a fractal environment is beneficial to our health. Freely walking individuals produce a multi-fractal time series, i.e. dynamics which vary among a range of fractal dimensions. If an external constraint in the form of a pacemaker is imposed on an individual's walk, an increase in randomness can be observed. If this constraint is a metronome, which produces a highly regular rhythm, the time series shows an increase in randomness and a reduction in long-term memory. West found that this was true for walkers with slow, normal and fast gait controlled by an allometric control (a metronome). He concluded that these individuals' response to stress was a loss of long-term correlations of long-term memory [14]. A loss of internal long-term correlations as a result of a reduction or loss in environmental long-term correlations severely cuts into a system's ability to display global strong anticipation. As we saw in Chapters 2 and 5, global anticipation emerges when internal and external long-term correlations, such as those generated by $1/f$ noise, are coordinated.

The relation between embodied and embedding constraints can be expressed in terms of their respective degree of complexity, measured in the fractal dimension. Elsewhere, I have denoted this relation as "interface complexity" [15]. Once the interfacial cut between the observant and environment has been determined by the assignment conditions [16], which determine what belongs to the observant and what to the environment, internal and external complexity can be assessed.

Interface complexity (IC) compares the complexity of embodied and embedding structures in terms of the number of nestings available, measured in Δt_{depth} and $\Delta t_{density}$. These can be defined as the range of possible responses an observant has at his or her disposal compared to the range of relevant environmental parameters. (This relationship is reciprocal, as the environment's range of responses also adapts to the observant's range of relevant parameters, in particular if the environment is another human being.)

Interface complexity is measured in Δt_{depth} for local strong anticipation and in the fractal dimension $\Delta t_{density}$ for global strong anticipation:

For **local strong anticipation**, IC is measured as the relation between embodied and embedding parameters:

$$\text{IC} = \Delta t_{depth}\ (\text{observant})/\Delta t_{depth}\ (\text{environment})$$

where the environment can also consist of another observant.

Examples of embodied and embedding parameters in direct coupling scenarios, as in anticipatory synchronization, are the frequencies of master and slave, measured in Hz.

For **global strong anticipation**, IC is measured in the fractal dimension as the relation between embodied and embedding long-term correlations:

$$\text{IC} = \Delta t_{density}\ (\text{observant})/\Delta t_{density}\ (\text{environment})$$

where the environment can also consist of another observant.

Examples of relevant parameters are HRV, SRV and external pacemakers in West's description of fractal time series but also examples with differing parameters, like leg strength and leg weight in Thelen and Smith's analysis of stepping behaviour.

9.2 Contextualization Generates Δt_{depth} and $\Delta t_{\text{density}}$, Long-Term Correlations and Global Strong Anticipation

If embedding parameters cause stress for the embedded obserpants, their response is a loss of long-term correlations of long-term memory. Stress can be induced by allometric pacemakers and also by a lack of external stimuli, as experienced in an isolation tank. Although gravity cannot be removed as an external parameter, environmental Δt_{depth} and $\Delta t_{\text{density}}$ are significantly reduced for an obserpant embedded in such a tank. As a result of lacking stimuli, and thus boiling in his or her own broth, such an obserpant generates hallucinations (see Fig. 9.1). In other words, such an obserpant creates an internal context which generates internal Δt_{depth} and $\Delta t_{\text{density}}$. Consequently, IC is primarily determined by the obserpant, as the environment has little impact.

When the values of parameters change over time, as in Thelen's and Smith's example of a baby's stepping behaviour, IC is modified accordingly: When leg strength increases, the embodied component of IC gains in relevance, and when leg weight increases, the embedding environmental component dominates IC.

Contextualization means an increase in nested simultaneity, as Δt_{depth} increases with every new embedding, both internally and externally. Taking account of both successive and simultaneous structures allows us to calculate the fractal dimension as $\Delta t_{\text{density}}$. $\Delta t_{\text{density}}$ includes succession and therefore describes long-term correlations, both internal and external, in terms of their fractal dimensions.

$1/f$ noise (also referred to as flicker noise or pink noise) is such a long-term correlation: a scale-invariant temporal extension which displays correlations between fluctuations on nested time scales. Van Orden *et al.* stress the importance of $1/f$ (or pink) noise for human beings:

> Pink noise is a fundamentally complex phenomenon that reflects an optimal coordination among the components of person and task environment. Departures from this optimum occur in advanced aging and dynamical disease ... [17]

Fig. 9.1. Francisco de Goya: The sleep of reason produces monsters [18].

As we saw earlier in this chapter, advanced age and dynamical diseases are often accompanied by deviations towards more structured or more random fluctuations (see also Chapter 11 on anticipation in dynamical diseases). Van Orden *et al.* denote successful coordination of embodied and embedding fluctuations as being "in the pink" — a feeling of well-being resulting from optimal fractal embedding. This is what Mihály Csíkszentmihalyi denoted as *flow*: successful synchronization between embodied individuals and their embedding task environment [19]. A promising candidate for that shared complexity is 1/*f* scaling. This is not surprising, as $1/f$ noise is ubiquitous in nature and has been shown to correlate with healthy dynamics and successful communication [20].

The coordination of internal and external $1/f$ long-range correlations is a likely underlying cause of global strong anticipation. The fact that many natural systems display fluctuations on nested time scales may be a selection effect. Stephen and Dixon therefore suggested that global strong anticipation may also be ubiquitous [21].

Contextualization generates Δt_{depth} and $\Delta t_{\text{density}}$, internal and external fractal long-term correlations and, if successfully coordinated, global strong anticipation. But although contextualization and living in the pink appear to be the healthy and successful options, the opposite mechanism — decontextualization — also offers advantages and healthy choices.

9.3 Decontextualization Reduces Δt_{depth} and $\Delta t_{\text{density}}$, Long-Term Correlations and Global Strong Anticipation

At the beginning of this chapter, we differentiated between two types of obserpants: fractal and nonfractal. Fractal obserpants have at their disposal a multi-level range of possible responses to a nested environment. Nonfractal ones lack this internal nested structure or display only a limited number of nested temporal levels. This reduces their capacity to create, recognize and coordinate with context. Reducing context is a two-edged sword: On the one hand, obserpants who fail to use context fall for auditory or visual illusions, but, on the other hand, this inability can also turn out to be a blessing in disguise.

One example of the failure to use context is the Shepard scale, an auditory illusion designed by psychologist Roger Shepard (see Fig. 9.2) [22]. This auditory illusion is created if all notes separated by an octave are played simultaneously in a continuous loop, which triggers the illusion of an ever-ascending tone. The illusion occurs because the listener focuses only on pitch relations, i.e. the frequency of the successive notes.

The same illusion occurs in the Risset Scale, a continuous version of the discrete Shepard Scale designed by composer Jean-Claude Risset [24].

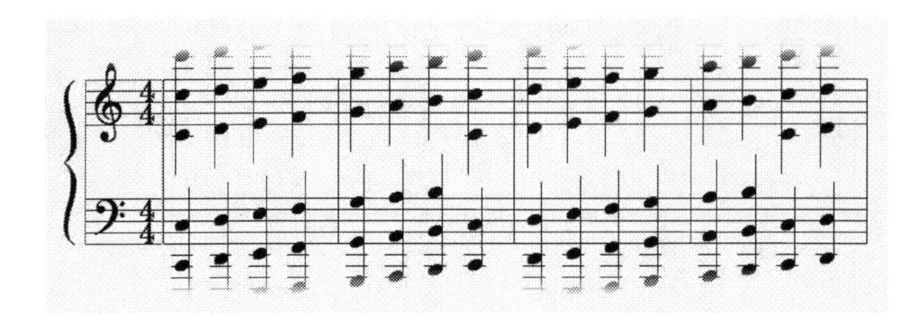

Fig. 9.2. The Shepard Scale [23].

Risset's endless scale illusion has been used in a number of musical pieces, for instance, in the Beatles' *A Day in the Life* [25].

In both the Shepard and the Risset scales, the illusion results from the listener's attempt to extract a one-dimensional signal from a multi-layered one. Focussing on the pitch only means perceiving pitch relations in Δt_{length} only, rather than those in Δt_{depth}. Such a limited perspective lacks nested simultaneity.

Against the background of a fractal, extended Now [26], a focus on succession rather than simultaneity leads to illusory perceptions because protensions are constrained by a one-dimensional signal. Such protensions are predictions without depth, i.e. they are not embedded in the signal's multi-dimensional framework. This is an example of ignoring context, of a decontextualization which generates an illusion.

As I have described in Chapter 9, there are also examples in which a failure to use context turns out to be a blessing in disguise. Neuroscientist Stephen Dakin *et al.* showed that individuals with a schizophrenic disorder perform better in a matching task in which a simultaneous contrast tends to compromise the perception of healthy subjects [27]. Both healthy and schizophrenic subjects were asked to look at the image of a high-contrast disk, inside which a smaller, low-contrast disk was embedded. Around the high-contrast disk, there were eight low-contrast disks of the same size as the embedded one but with slightly varying low contrasts. The task consisted of matching one of the outer disks with the embedded one in terms of their contrast. The schizophrenics concentrated on each disk individually, rather than looking at them from a global perspective in which all disks contextualize each other. In other words, they studied them successively, i.e. in Δt_{length}. By contrast, healthy subjects studied them simultaneously, i.e. in the temporal dimension of Δt_{depth}. As a result, they fell for the simultaneity contrast which distorted the contrast of the embedded disk and failed to match the correct smaller disk with the embedded one. The inability of the schizophrenic subjects to contextualize the inner disk and thus generate in Δt_{depth} prevented them from falling for the optical illusion.

Other advantages of reducing or eliminating context show themselves in stressful situations which require immediate focus on relevant stimuli only. Tunnel vision helps individuals to concentrate in a life-threatening situation by eliminating all distracting stimuli. An example is a law enforcement officer in a shooting incident, who reported experiencing tunnel vision, during which he did not hear gunshots nor the utterances of bystanders, and his visual field was reduced to the target. Time slowed down for him because he had reduced context to an absolute minimum and thus did not generate in Δt_{depth}. As a result, Δt_{length} increased in his temporal perspective, expanding his Now. Thus, tunnel vision is an example of potentially life-saving decontextualization [28].

Ridding themselves of contexts enables obserpants to focus also in less dramatic situations, for instance, in meditation, when the meditating individual concentrates on succession only and tries to eliminate all distracting simultaneous stimuli or thoughts. Like in tunnel vision, Δt_{depth} is reduced to an absolute minimum, while Δt_{length} increases.

In general, an increase in Δt_{depth} means that more context exists which generates a multi-layered, fractal Now with nested retensions and protensions. A decrease in Δt_{depth} reduces not only the number of simultaneous layers but also the fractal dimension of the Now, which is measured in $\Delta t_{\text{density}}$. As a result, the focus lies on Δt_{length}, i.e. on mere succession of almost uncorrelated perceptions of external stimuli. (There are always some embeddings in Δt_{depth} which we cannot remove, such as the impact of gravity or internal bodily oscillations, which guarantees that there will always be some correlations.) Thus, a decrease in $\Delta t_{\text{density}}$ removes long-term correlations, which are a necessary condition for global strong anticipation.

References

[1] J. L. Casti, *Complexification — Explaining a Paradoxical World through the Science of Surprise*, HarperPerennial, New York, 1995, pp. 276–269.

[2] *ibid*, pp. 276–277.

[3] R. Buccheri, Intelligibility, endophysics and time. In: R. Buccheri, M. Saniga and W. M. Stuckey (eds.) *The Nature of Time: Geometry, Physics and Perception*, NATO Science Series, Vol. 95, Kluwer Academic Publishers, 2003, p. 414.

[4] E. Thelen and L. B. Smith, *A Dynamic Systems Approach to the Development of Cognition and Action*, Bradford Books/MIT Press, Cambridge, Mass., 1994; S. Vrobel, Spanning and stretching of temporal fields. G. E. Lasker (ed.) *Acta Systemica* (IIAS, Tecumseh, Canada), XIII(1), 9–14, 2013.

[5] G. C. van Orden, H. Kloos and S. Wallot, Living in the pink: Intentionality, wellness and complexity. In: C. Hooker (ed.) *Philosophy of Complex Systems: Handbook of the Philosophy of Science*, Elsevier, Amsterdam, 2009; E. Thelen and L. B. Smith, *A Dynamic Systems Approach to the Development of Cognition and Action*, Bradford Books/MIT Press, Cambridge, Mass., 1994.

[6] E. Thelen and L. B. Smith, *A Dynamic Systems Approach to the Development of Cognition and Action*, Bradford Books/MIT Press, Cambridge, Mass., 1994.

[7] B. J. West, *Where Medicine Went Wrong — Rediscovering the Path to Complexity*, World Scientific, Singapore, 2006.

[8] *ibid*, pp. 283–284.

[9] *ibid*, p. 176.

[10] *ibid*, pp. 288–289.

[11] *ibid*, p. 303.

[12] *ibid*, p. 219.

[13] *ibid*, p. 221.

[14] *ibid*, p. 305.

[15] S. Vrobel, Nesting performances generate simultaneity: Towards a definition of interface complexity. In: Trappl, R. (ed.) *Cybernetics and Systems. Proceedings of the European Meeting on Cybernetics and Systems Research*, Vol. 2, Austrian Society for Cybernetics Studies, Vienna, Austria, 2006, pp. 375–380.

[16] O. E. Rössler, *Endophysics*. World Scientific, Singapore, 1998.

[17] G. C. van Orden, H. Kloos, and S. Wallot, Living in the pink: Intentionality, wellness and complexity. In: C. Hooker (ed.) *Philosophy of Complex Systems: Handbook of the Philosophy of Science*, Elsevier, Amsterdam, 2009, p. 639.

[18] F. de Goya, *El sueño de la razón produce monstruos* (The sleep of reason produces monsters), 1797–1799, Capricho Nr. 43, Aquatinta-Radierung, 21,6x15,2 cm, Museo de Calcografia, Madrid (Public domain).

[19] M. Csíkszentmihalyi, *Flow — The Psychology of Optimal Experience*, Harper Perennial, New York, 1990.

[20] V. Marmelat and D. Delignières, Strong anticipation: Complexity matching in inter-personal coordination, *Experimental Brain Research*, 222, 137–148, 2012; D. G. Stephen and J. A. Dixon, Multifractal cascade dynamics modulate scaling in synchronization behaviours. *Chaos, Solitons & Fractals* (Elsevier), 44(1–3), 160–168, 2011; G. van Orden, H. Kloos und S. Wallot, Living in the pink: Intentionality, wellbeing, and complexity. In: C. Hooker (eds.) *Philosophy of Complex Systems, Handbook of the Philosophy of Science*, Elsevier, Amsterdam, 2011, pp. 629–672; B. J. West, *Where Medicine Went Wrong — Rediscovering the Path to Complexity*, World Scientific, Singapore, 2006, pp. 283–284.

[21] D. G. Stephen and J. A. Dixon, Multifractal cascade dynamics modulate scaling in synchronization behaviours. *Chaos, Solitons & Fractals* (Elsevier), 44(1–3), 160–168, 2011.

[22] R. N. Shepard, Circularity in judgments of relative pitch. *Journal of the Acoustical Society of America*, 36, 2346–2353, 1964.

[23] Reprinted with permission of Nicola Titeux. https://www.nicolastiteux.com/en/blog/shepard-and-risset-audio-illusions/.

[24] J.-C. Risset, Pitch control and pitch paradoxes demonstrated with computer-synthesized sounds. Journal of the Acoustical Society of America, 46, 88, 1969.

[25] The Beatles, 1967/2009, track 13; E. Vernooij, A. Orcalli, F. Fabbro and C. Crescentini: Listening to the Shepard-Risset Glissando: The relationship between emotional response, disruption of equilibrium, and personality. *Frontiers in Psychology, Cognition*, 7, 2016.

[26] S. Vrobel, Temporal observer perspectives. G. E. Lasker (ed.) *IIAS-Transactions on Systems Research and Cybernetics* (Tecumseh, Canada), VII(1), 1–10, 2007.

[27] S. Dakin e*t al.*, Weak suppression of visual context in chronic schizophrenia. *Current Biology*, 15, R822–R824, 2005.

[28] A. Artwohl, Perceptual and memory distortion during officer-involved shootings. J. E. Ott (ed.) *FBI Law Enforcement Bulletin* (FBI, Pennsylvania Ave., Washington, DC), 71(10) , October 2002; S. Vrobel, When time slows down: The Joys and Woes of De-Nesting. George E. Lasker (ed.) *Acta Systemica* (IIAS, Tecumseh, ON, Canada), X(1), 2010, pp. 33–40; S. Vrobel, *Fractal Time — Why a Watched Kettle Never Boils,* World Scientific, Singapore, 2011.

Chapter 10

Compensated Delays as Phenomenal Blind Spots

We are not always aware of the simultaneous contrasts we are exposed to. Neither are we conscious of the fact that compensated delays and distances are examples of local strong anticipation at work. The transparency which results from compensatory acts enables us to skip disadvantageous delays and allows us to navigate the world smoothly. However, local strong anticipation is also a blind spot, as it conceals hidden compensatory acts. As such, it presents a constraint to any epistemological endeavour. As Metzinger put it, only if "darkness is made explicit", i.e. phenomenal transparency is made visible, can we reveal compensated temporal and spatial extensions we did not even suspect existed.

This chapter deals with compensated delays as phenomenal blind spots and demonstrates a new kind of relativity among obserpant types. Endo- and exo-perspectives differ when it comes to perceiving compensated delays, even temporal order is perceived differently during recalibration. Against this background, examples of local strong anticipation are interpreted as blind spots which may evolve into natural laws.

10.1 Compensated Delays Are Phenomenal Blind Spots

Phenomenologist Maurice Merleau-Ponty described our bodies as situated and extended in a spatio-temporal environment. He insisted that the Cartesian Cut was a misleading concept, as both *res cogitans* and *res extensa* are reconciled in one bodily existence. And as our perceptual apparatus is embedded in a spatio-temporal environment, we are part of the system we wish to describe and our bodies thus necessarily form a blind spot of perception. Merleau-Ponty describes the relationship between the environment and a human body, which is both limited by and influences its environment through its physical orientation in space and its rhythm:

> The presence and the absence of external objects are only variations within a field of primordial presence, a perceptual domain over which my body has power. (...) If objects must never show me more than one of their sides, then this is because I myself am in a certain place from which I see them, but which I cannot see. If I nevertheless believe in their hidden sides, as well as in a world that encompasses them all and that coexists with them, I do so insofar as my body, always present for me and yet engaged with them through so many objective relations, maintains them as coexisting with it and makes the pulse of its duration reverberate through them all. [1]

Merleau-Ponty describes our Now — as did Edmund Husserl — as an extended temporal perspective which hosts both succession and nested simultaneity. As a starting point, he envisages a "densely embodied", extended "Here-Now", which stretches both into the past and the future and is transparent to us ("time only has sense for us because we 'are it'") [2]. In other words, our Here-Now is our phenomenal blind spot:

> Time only exists for me because I am situated in it, that is, because I discover myself already engaged in it, (...) because a sector of being is so close to me that it does not even sketch out a scene in front of me and because I cannot see it, just as I cannot see my own face. [3]

In Merleau-Ponty's terminology, human beings "imbue" and "impregnate" their environment with their physical and temporal extensions. He thus describes a two-way relationship between the embedding environment and the embedded obserpant, which together form a systemic whole. Like Bateson's *unit of survival*, the systemic whole is a flexible organism situated in a flexible environment [4].

This bidirectional coupling of obserpant and environment results from an interfacial shift between originally two systems. The interfacial cut between the original obserpant and environment was shifted to the outside so as to assimilate parts of the environment in the form of distances or delays. As we saw in earlier chapters, the compensation of external delays reduces our interface complexity, whereas uncompensated delays increase it. If the obserpant displays a higher degree of complexity than the environment, this manifests itself in a wider range of responses on the part of the obserpant.

If the environment turns out to be more complex than the obserpant, the latter reduces that higher degree of complexity through compensatory acts. This can be achieved either through symbol systems, as in social trust formation or embodied complexity reduction through delay compensation [5]. Both types of compensation are transparent to the extended obserpants, as they are not aware that they have reduced their environmental complexity — or, in the Merleau-Pontian terminology — that such compensations are blind spots.

10.2 Recalibration

We are not aware of compensated delays. This is true for postdiction of neurocognitive delays which result from different transmission and processing times of sensory signals. The fact that our brain waits for the slowest signal to arrive and then integrates the multi-modal signals into a coherent experience is transparent to us [6]. So is the window of delay which we compensate postdictively — it is a phenomenal blind spot.

The same is true for sensory-motor delays. Here, however, delay compensation does not result from postdiction but from local strong anticipation. When we play tennis, we don't aim our racket at the current, "real-time", position of the ball, but intercept the trajectory at a point the ball will reach after a few milliseconds. Those milliseconds also present a phenomenal blind spot. Both delay compensation through postdiction and through local strong anticipation are transparent to the obserpant. Local strong anticipation manifests itself as a blind spot to the obserpant.

Such blind spots emerge when the obserpants need to modify their behaviour because they have to adapt to environmental modifications. This happens when expectations are not met and the obserpant's hypotheses about the world prove to be wrong. Then the working hypothesis needs to be updated by recalibrating, i.e. by adjusting to delays or other environmental changes (such as spatial distance modifications). Once compensated, the adjustment becomes a phenomenal blind spot.

Not only temporal delays require compensation and recalibration. Spatial obserpant extensions such as the stick for reaching in Berti and Frassinetti's experiment (see Chapter 3) also undergo a recalibration process before the spatial "gap" is compensated. The compensation becomes transparent to the user when the stick becomes embodied as an elongated limb, i.e. when a new systemic whole is created.

Stetson *et al.*'s keypressing experiments show how fast obserpants adapt to recalibration after the insertion and removal of an existing delay. After a brief interval of disorientation and confusion of causal relations, the inserted or removed delay is compensated and the obserpant navigates his or her environment again smoothly and with confidence [7]. For an external obserpant, however, the witnessed recalibration and delay compensation are not transparent.

10.3 A New Kind of Relativity: Comparing Compensated and Uncompensated Perspectives

Although transparency appears to be part of our *condition humaine*, it is not impossible to make some hidden interfaces visible. However, this requires an additional perspective, preferably that of another obserpant. Ideally, this external obserpant displays a different degree of internal complexity, i.e. a different range of possible responses. This facilitates making the interfacial cuts visible.

The measure of choice to compare endo- and exo-perspectives is the distribution of Δt_{depth} and Δt_{length}. Inserted delays, which are visible to an external observant, are transparent from the endo-perspective once the delay has been compensated. An external observant, by contrast, will not be subjected to confusion resulting from recalibration, as he or she can clearly make out the interfacial cut and identify the compensation [8].

We are embedded in and embed compensated and uncompensated delays in our social, biological and physical extensions and environments. If one observant has compensated a delay which for another remains uncompensated, they may disagree whether a temporal relation is simultaneous or successive.

In the following comparisons of endo- and exo-perspectives, two are examples of delay/distance compensation and thus of local strong anticipation, while one portrays a loss of local strong anticipation (and utter confusion and surprise because an existing delay has been removed).

Tables 10.1 and 10.2 compare the endo- and exo-perspectives in Stetson *et al.*'s keypressing experiments. Table 10.1 shows that if a delay is inserted into a control loop (as in their first experiment), the endo- and exo-perspectives differ after recalibration and delay compensation. After a delay is inserted, the observant's endo-perspective shows an increase in Δt_{length}, while Δt_{depth} remains unchanged. The same is true for the external observant. Once the delay has been compensated, endo- and exo-perspectives differ: Now Δt_{depth} has increased from the endo-perspective, while Δt_{length} has decreased. By contrast, from the exo-perspective, both temporal

Table 10.1. If a delay is inserted, endo- and exo-perspectives differ after recalibration and delay compensation. From the endo-perspective, Δt_{depth} has increased while Δt_{length} has decreased. An external observant experiences no change in perceived temporal dimensions [9].

	Observant (Endo-perspective)		**External observant (Exo-perspective)**	
	$y = \Delta t_{depth}$	$x = \Delta t_{length}$	$y = \Delta t_{depth}$	$x = \Delta t_{length}$
After delay insertion	y	$x + 1$	y	$x + 1$
After delay compensation	$y + 1$	x	y	$x + 1$

Table 10.2. If an existing delay is removed, endo- and exo-perspectives differ both before and during recalibration. After the removal of the existing delay, Δt_{length} has decreased from the endo-perspective, which results in an apparent causal reversal. An external obserpant is aware first of the existing and then of the removed delay and experiences no change in perceived temporal dimensions after the removal of the delay.

	Obserpant (Endo-perspective)		**External obserpant (Exo-perspective)**	
	$y = \Delta t_{\text{depth}}$	$x = \Delta t_{\text{length}}$	$y = \Delta t_{\text{depth}}$	$x = \Delta t_{\text{length}}$
Before removal of an existing delay	$y + 1$	x	y	$x + 1$
After removal, during recalibration	y	$x - 1$	y	x
After recalibration	y	x	y	x

dimensions remain unchanged. The endo-obserpant exhibits local strong anticipation after the delay compensation, whereas no anticipative regulation takes place from the exo-perspective.

(In Tables 10.1, 10.2 and 10.3, it is assumed that the external obserpant neither compensates nor recalibrates the delay/advance. Think of this obserpant as an inanimate monitoring device.)

In the second setup of Stetson *et al.*'s experiments, an existing delay was removed. Table 10.2 shows that both before and after the removal, the endo- and exo-perspectives differ. Although the obserpant is not aware of the increase in Δt_{depth} resulting from the previous compensation of an inserted delay, Δt_{depth} has since then increased. From an external perspective, the previously added delay is still registered (as the exo-obserpant does not recalibrate), so Δt_{length} has still increased. After the removal, during recalibration, the endo-perspective shows a decrease in Δt_{length}, which results in an apparent causal reversal. An external obserpant is aware of the previously existing delay and experiences no change in perceived temporal dimensions after the removal of this delay. The endo-obserpant no longer exhibits the local strong anticipation acquired after the delay compensation in the first experiment (neither does the exo-obserpant).

Table 10.3. If a tool is incorporated, endo- and exo-perspectives differ after recalibration and distance (strictly speaking, spatio-temporal) compensation. From the endo-perspective, Δt_{depth} has increased while Δt_{length} has decreased. An external obserpant experiences no change in perceived temporal dimensions.

	Observant (Endo-perspective)		**External obserpant (Exo-perspective)**	
	$y = \Delta t_{depth}$	$x = \Delta t_{length}$	$y = \Delta t_{depth}$	$x = \Delta t_{length}$
Before incorporation of tool	y	$x + 1$	y	$x + 1$
After tool incorporation and recalibration (= distance compensation)	$y + 1$	x	y	$x + 1$

Like delay compensation, compensated spatial distances also lead to local strong anticipation. Transforming extrapersonal space into peripersonal space is an example of such distance compensation (which manifests itself as spatio-temporal compensation).

Table 10.3 shows an increase in Δt_{length}, as long as the systemic-whole-to-be consists of arm + stick. After recalibration, arm + stick have merged into one elongated limb, whose components are moved simultaneously as one systemic whole. After this distance compensation, the endo-obserpant exhibits local strong anticipation, while the exo-obserpant registers no change.

To compare the endo-perspectives in terms of $\Delta t_{density}$, say, in the perception of overtones with removed fundamentals, the fractal dimension can be calculated by assigning n to the number of sinus waves in Δt_{length} and s to the scaling factor. The resulting similarity dimension equals $D = \log(n)/\log(s)$.

To compare the range of responses, i.e. internal complexity in terms of the fractal dimension perceived, one may compare the ability to perceive a missing fundamental in different obserpant types. Infants younger than 3 months cannot perceive the missing fundamental, while after 4 months, it is integrated. Developmental psychologists Chao He and Laurel Trainor conducted experiments on the ability to perceive the missing fundamental and found that

> what is clear is that between 3 and 4 months of age there is a major shift in how pitch is represented in the cortex, such that by 4 months components that stand in harmonic relations fuse into a single percept whose pitch corresponds to the fundamental, whether or not it is actually present in the stimulus. [10]

For obserpants younger than 3 months, there is no integration of overtones into a fractal percept. So, in terms of fractal time, Δt_{depth} and, consequently, $\Delta t_{\text{density}}$ remain unchanged. By contrast, obserpants of 4 months and older do fuse the missing fundamental and the overtones into a fractal percept. To them, Δt_{depth} and $\Delta t_{\text{density}}$ increase, which means they have developed local strong anticipation.

10.4 Could Local Strong Anticipation Evolve into Natural Laws?

Embedded and embedding factors determine interface complexity and the degree of local strong anticipation in obserpants. We have seen that local strong anticipation in the form of delay and distance compensation is a means of forming habits, which change our interfacial structures, often with corresponding changes on the neural level. Elsewhere, I raised the question as to whether we may conclude that the habit-forming property of local strong anticipation brings forth natural laws [11].

Natural laws may seem as if they don't require explanation, but this could be so because the explanation is transparent to us. Jesper Hoffmeyer suggested that we should follow Charles Sanders Peirce's recommendation and treat natural laws as phenomena which themselves necessitate explanation [12]:

> Much in the same way as Einstein recontextualized the 'universal' laws of Newton by showing them to be the local products of more general principles, Peirce saw natural laws as the secondary products of a more general tendency in the universe to generate regularities (or *habits* as he often called them). Rather than seeing the universe as characterized by lawfulness, its primary state, according to Peirce, is indeterminacy and chance. And thus the formation of regularities, such as for instance

> natural laws, must be explained by other means — for natural law is thus a product of evolution, not its source. [13]

Hoffmeyer suggests that the generation of regularities in the universe evolves into semiosis, our ability to form *interpretants* — self-perpetuating habits.

Are our compensatory acts, which manifest themselves as local strong anticipation, blind spots which may evolve into natural laws? As such, they would evade our awareness. Successful compensation is transparent, and as phenomenal and epistemological blind spots, compensatory acts result in constraints on the obserpant. However, making them visible, or, in Metzinger's terms, opaque (i.e. not transparent) is not an impossible task, as we have seen in experiments which inserted delays or removed existing ones. If we see local strong anticipation as a natural constraint, an evolved regularity which, to the obserpant, has become a habit, it has, from the endo-perspective, acquired the status of a natural law. For the exo-obserpant, habits and regularities which evolved for the endo-obserpant will merely have the status of assignment conditions in Rössler's sense (see Section 2.3). However, intersubjectively emerging regularities can be upgraded into natural laws that hold true for a group of obserpants or other reference systems. We have tools to reveal the existence of compensated and uncompensated delays and distances, the relativity of fractal and nonfractal obserpants, and the fact that strong anticipation is a blind spot in our perception. By comparing obserpant types, we can become aware of discrepancies in their perceptions, as described in the previous section.

We have almost certainly compensated temporal and spatial extensions we do not know exist and which manifest themselves as habits and regularities that act as natural constraints and laws outside our current awareness. We can draw analogies and suspect that more such blind spots in the form of local strong anticipation limit our perception of and interaction with the world. In order to reveal their existence, we need to compare endo- and exo-perspectives. As Laplace's demon is not available as an external observer, we shall have to focus on comparing endo-perspectives, preferably those of obserpants with differing fractal dimensions. What is transparent from one perspective may be visible for another obserpant.

At the moment, nobody can claim to know the extensions of local strong anticipation. My guess is that we are only aware of a tiny proportion of all the neurocognitive, sensory-motor and environmental delays humans are subjected to and have compensated. Revealing those hidden compensated delays which are characteristic of our human condition will be a promising endeavour. We may stumble upon a compensated delay or distance when we are surprised to find that others have not anticipated it. Or we may find that the cause for the delay is still present, but we no longer compensate it.

We may have to question how to classify "useful" local strong anticipative behaviour, as compensatory acts can be both advantageous and disadvantageous. The following chapter discusses the role of compensated blind spots in medical conditions. Dynamical diseases and disorders are presented against the background of healthy and pathological individuals' strong anticipative faculties.

References

[1] M. Merleau-Ponty, *Phenomenology of Perception*, Routledge, New York, 2012 (1945), pp. 94–95.

[2] *ibid*, p. 494.

[3] *ibid*, p. 447.

[4] G. Bateson, *Steps to an Ecology of Mind*, Chandler Publishing Company, Toronto, 1972, p. 451; S. Vrobel, A new kind of relativity: Compensated delays as phenomenal blind spots. *Progress in Biophysics and Molecular Biology* (Elsevier), 2015.

[5] N. Luhmann, *Vertrauen,* 4th edition, Lucius & Lucius, Stuttgart, 2005 (First published 1968); D. M. Eagleman, *Brain Time*. www.edge.org/3rdculture/eagleman09/eagleman09index.html, 20.10.2009; A. Noë, *Action in Perception*, MIT Press, Cambridge, 2004; A. Clark, *Supersizing the Mind. Embodiment, Action and Cognitive Extension*, Oxford University Press, New York, 2008; see also Chapter 7.

[6] D. M. Eagleman, *Brain Time*. www.edge.org/3rdculture/eagleman09/eagleman09index.html, 20.10.2009; see also Chapter 3.

[7] C. Stetson, X. Cui, P. R. Montague and D. M. Eagleman, Illusory reversal of action and effect. *Journal of Vision*, 5, 769a, 2005.

[8] S. Vrobel, A new kind of relativity: Compensated delays as phenomenal blind spots. *Progress in Biophysics and Molecular Biology* (Elsevier), 2015.
[9] A cone model of the relations presented in this table is published in S. Vrobel, A new kind of relativity: Compensated delays as phenomenal blind spots. *Progress in Biophysics and Molecular Biology* (Elsevier), 2015.
[10] C. He and L. J. Trainor, Finding the pitch of the missing fundamental in infants. *The Journal of Neuroscience*, 29(24), 7721, 17 June 2009.
[11] S. Vrobel, A new kind of relativity: Compensated delays as phenomenal blind spots. *Progress in Biophysics & Molecular Biology* (Elsevier, London), 119, 303–312, 2015.
[12] C. S. Peirce, 1992. In: K. L. Ketner (ed.) *Reasoning and the Logic of Things. The Cambridge Conferences Lectures of 1898*, Harvard University Press, Cambridge, MA, 1992; J. Hoffmeyer, *Biosemiotics: An Examination into the Signs of Life and the Life of Signs*, University of Scranton Press, London, 2008.
[13] J. Hoffmeyer, *Biosemiotics: An Examination into the Signs of Life and the Life of Signs*, University of Scranton Press, London, 2008, p. 95.

Chapter 11

Strong Anticipation as an Indicator of Well-Being

Both local and global strong anticipation shift an individual's or artificial agent's interfacial cut towards the outside, thus extending their sphere of influence. Local strong anticipation can result from extended obserpants compensating spatio-temporal extensions and also from synchronizing with two or more environmental systems. Global strong anticipation results from the coordination of long-term correlations of fluctuations on multiple time scales on both sides of the interfacial cut.

Local strong anticipation through delay and distance compensation by extended obserpants generates interfacial blind spots and usually has a positive impact on physical and mental health. By contrast, local synchronization through anticipatory regulation or direct coupling can lead to both healthy and pathological states.

Global strong anticipation, on the other hand, is generally regarded as being conducive to good health. Pink noise ($1/f$ noise), which exhibits fractal long-term correlations, has been observed in healthy dynamics and seems to be the basis for global strong anticipation. It represents the middle path between predictable pathological rigidity as in Brownian noise and a loss of structure in unpredictable white noise. Deviations from pink noise tend to cause pathological states.

If contextualization, i.e. the formation of Δt_{depth}, is compromised, fewer nestings result, which adds up to a lower fractal dimension. This, in turn, means that the obserpant's interfacial structure becomes less complex, so his or her range of possible responses is diminished. A shift in the temporal dimension from Δt_{depth} to Δt_{length} can be an indicator of dynamical diseases or disorders.

When I muse about the impact of temporal dimensions on health and well-being, Kurt Vonnegut's character Billy Pilgrim, the involuntary time traveller, invariably comes to mind. Billy's favourite author wrote a book about the demise of the Earthling's temporal constraints:

> It was about people whose mental diseases couldn't be treated because the causes of the diseases were all in the fourth dimension, and three-dimensional Earthling doctors couldn't see those causes at all, or even imagine them. [1]

This chapter investigates the impact of modifications in the temporal dimensions Δt_{depth} and $\Delta t_{\text{density}}$ on our anticipatory faculties and our well-being.

11.1 In the Pink

In Chapter 5, pink noise was introduced, along with white and Brownian noise, as a statistical temporal fractal, which means that short-term and long-term variations are statistically self-similar. While pink noise is ubiquitous in nature, its origin is not known [2]. We saw that long-term correlations in the form of pink noise exist in both organism and environment (including other organisms) and that coordination between such internal and external long-term correlations is conducive to communication and well-being.

Guy van Orden *et al.* presented the positive characteristics of pink noise in their seminal chapter "Living in the Pink". As briefly mentioned in Chapter 9, pink noise appears to be the ideal match between the internal dynamics of a person and that of his or her environment [3].

Van Orden *et al.* stress that in complexity science, constraints are to be understood as ephemeral structures created by temporary coupling in embodied and environmental components [4]. An example is Thelen's and

Smith's experiments with babies' stepping behaviour, which introduce the concepts of embodied and embedding constraints (see Chapter 9).

The fractal structure results partly from the fact that oscillations within an organism are linked to environmental rhythms and partly from the iterative processes which generate the time series. Oscillations which vary more slowly act as constraints to maintain oscillations which vary at a faster rate but not vice versa:

> ... the very slowly changing constraints limit the degrees of freedom available to a faster changing process, thus restricting the degrees of freedom for what can happen at faster time scales. The faster changing dynamics must evolve within the limited degrees of freedom that the context leaves available. [5]

As behavioural time scales are too slow to be controlled by the brain, van Orden *et al.* conclude that the "blue-collar" brain is controlled by the "white-collar" body [6]. Figure 11.1 shows the difference in the spans of time scales in behavioural and brain activities.

To visualize fractal behaviour, data can be portrayed in a spectral plot. Van Orden describes the method from scratch:

> To construct a spectral plot, begin by decomposing a time-ordered data series into sine waves of different amplitudes (see Fig. 11.2). Slow large changes in the data series are captured by the slow-frequency large-amplitude sine waves (top left of Fig. 11.2), and fast changes are captured by fast frequency small-amplitude waves (bottom left of Fig. 11.2). The amplitude or *power* (amplitude squared) concerns the size of particular changes S(f) and appears on the y-axis of the power spectrum. Size of change S(f) is plotted against the frequency (f) of changes, which is also an estimate of how often changes of that size occur (on log-log scales). The slope of the regression line between how often (f) and how big S(f) in the spectral plot estimates the scaling relation between size and frequency of change. In Fig. [11.2], the size of change S(f) is inversely proportional to its frequency (f): $S(f) = 1/f^{\alpha} = f^{-\alpha}$, with scaling exponent $\alpha \approx 1$, the scaling exponent of pink noise. [8]

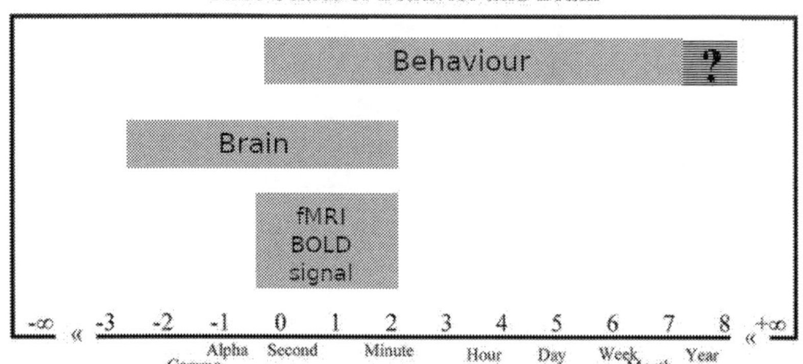

Fig. 11.1. Duration of Sine Wave Periods in $\log_{10}$(Sec), Approximating Empirical Variation: Time scales of behaviour and brain were estimated from the time scales implicated in sine wave simulations of variation across repeated measurements. Landmarks of durations (day, week, etc.) or brain activity (alpha and gamma) are placed near their values in log10 (Sec). This figure also includes the span of brain activity observed in the BOLD signal of brain metabolism used in fMRI studies, all to give context to the contrast between the span of time scales observed of behaviour and the span of time scales observed of the brain. The question mark to the right of the behavioural span symbolizes the fact that no upper bound short of death has yet been discovered in longitudinal studies estimating the presence of scaling relations in the variation across measurements of behaviour [7].

Van Orden *et al.* conclude that pink noise may be the ideal match between embodied and embedding constraints. As we saw in Chapter 9, pink noise correlates with successful communication and healthy dynamics, whereas white or Brownian noise reflects pathological dynamics in HRV and SRV. With a few notable exceptions, pink noise appears to signal a healthy middle path between randomness and stifling regularity.

11.2 Global Strong Anticipation as an Indicator of Health

A possible basis for global strong anticipation has been identified as the coordination between internal and external $1/f$ long-range correlations, i.e. pink noise [9]. Against this background, the ubiquity of pink noise led Stephen and Dixon to suggest that the coordination of internal and

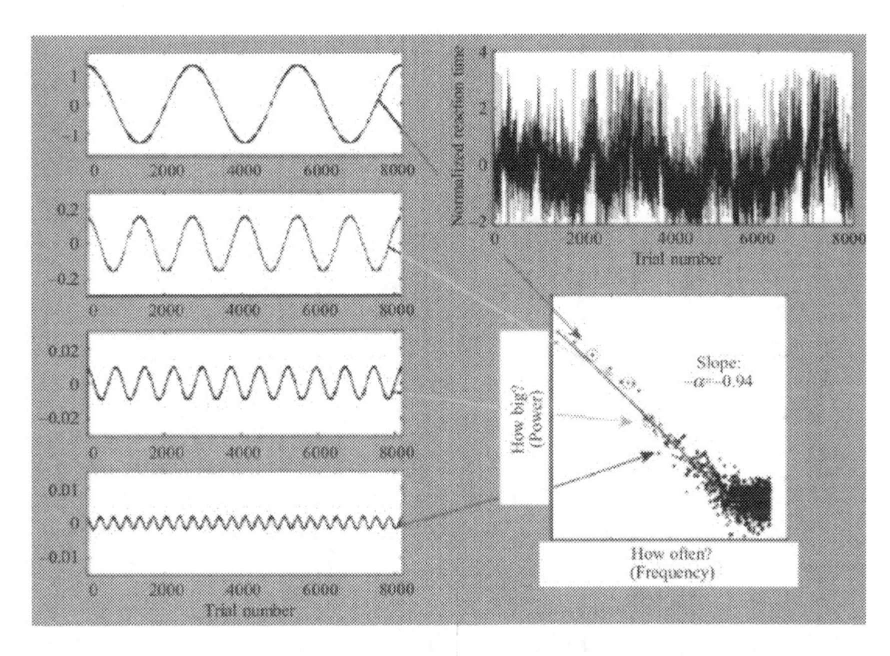

Fig. 11.2. One person's response time data [in a button-pressing experiment]. The left side of the figure presents specific frequencies and amplitudes of the regular sine waves that approximate the irregular aperiodic data series (in the upper right of the figure). The arrows connect the sine waves to corresponding points of the spectral plot. Each point plotted represents a particular size or amplitude of change (power) across the data values and how often changes of that size occur (frequency). The spectral slope = –0.94, which is approximately $\alpha \approx 1$. Note that the y-axes in the sine wave illustrations have been adjusted to make smaller amplitude sine waves visible [11].

external fluctuations on nested time scales, and thus global strong anticipation, is a selection effect [10].

Within the fractal perspective of the anticipating obserpant, nested time scales may manifest themselves in integrative processes in which a perceptual gestalt emerges. Such integration would not result in synchrony, i.e. congruence of the self-similar structures on all nested levels, but in a more complex gestalt which results from resonance (as is the case in the coordination of internal and external pink noise). By contrast, temporal congruence, i.e. synchrony, results in a loss of Δt_{depth}, as no overlapping nesting occurs. Nesting generates Δt_{depth} and thus a fractal structure, in which the self-similar structure that expands in $\Delta t_{\mathrm{depth}}$ occurs on all nested levels. The resulting fractal perspective creates an extended Now for the obserpant and thereby increases the range of possible responses to environmental stimuli and change. The coordination of internal and external long-term correlations in fluctuations (as in pink noise) widens the obserpant's temporal interface, the extended Now, and provides the basis for global strong anticipation.

Examples of pink noise as health-inducing structures can be found in many areas, such as successful communication and other forms of interpersonal coordination, healthy heartbeat, breathing, gait and digestion (see Chapter 5) and also in stress-reducing visual and auditory environments [12]. In music, for instance, compositions from Brahms to the Beatles resemble pink noise in the way the size of the intervals between successive musical notes varies. A typical tune would avoid very large and very small changes so that a balance between rigid order (as in Brownian noise) and uncorrelated unpredictability (as in white noise) emerges. This balance guarantees that the tune is neither too predictable (and thus boring) nor too erratic (which would overburden the listener with too much novelty) [13]. Music appears to imitate the characteristic way the world changes. And an environment in which embodied and embedding constraints display the same degree of complexity is most conducive to global strong anticipation.

Data scientist Summer Rankin *et al.* showed that pink noise also governs expressive tempo fluctuations in musical performances. For listeners to be able to anticipate the onset of changes in tempo, the structure of pink noise proved to be sufficient [14]. The researchers suggest that such temporal synchronization which allows listeners to anticipate events in non-isochronous rhythms may improve therapies for Morbus Parkinson and related disorders.

The stress-reducing effect of pink noise, be it in the form of music, ocean surf or nonauditory stimulation, is well documented. Pink noise facilitates stress relief by masking tinnitus, improves postural sway control and the quality of sleep and enhances default-mode network activity in individuals with early Alzheimer's disease [15]. It also improves medical devices by providing a fractal interface: Pink noise input allows stethoscopes to listen to multiple frequencies simultaneously to avoid signal distortion as generated with conventional pure tone sweep [16]. Our own inbuilt fractal interfaces make the human being a multi-level, fractal detector whose range of internal responses determines the quality of its coordination with its environment. It appears our body is an equation which reflects the balance between internal and external coherence [17].

11.3 Dynamical Diseases Can Result from a Reduction in or Loss of Δt_{depth} and Turning Δt_{depth} into Δt_{length}

As we saw in Chapter 5, deviations from pink noise toward Brownian or white noise are detrimental to health. A breakdown of long-term correlated multi-scale dynamics leads to pathological states. As both Bruce West and Ary Goldberger *et al.* emphasize, a breakdown of fractal long-term correlations results in either uncorrelated randomness or rigid order. Excessive order manifests itself in pathologic periodicity, which is a hallmark of Morbus Parkinson, epilepsy, obstructive sleep apnea, sudden cardiac death, fetal distress syndromes, etc. [18]. Goldberger *et al.*, among others, stress that the range of possible responses to unpredictable environmental stimuli is a defining feature of a healthy constitution. They attribute this ability to adapt to external perturbations to two factors:

> *(i)* long-range correlations serve as a (self) organizing mechanism for highly complex processes that generate fluctuations across a wide range of time scales; and *(ii)* the absence of a characteristic scale inhibits the emergence of highly periodic behaviours (mode-locking), which would greatly narrow functional responsiveness. The latter conjecture is supported by findings from life-threatening conditions such as heart failure, where the breakdown of fractal correlations is often accompanied by the emergence of a dominant mode, e.g. the Cheyne-Stokes breathing frequency. [19]

Such mode-locking is usually prevented. In the brain, for instance, phases do not easily synchronize:

> A critical aspect of brain oscillators is that the mean frequencies of the neighbouring oscillatory families are not integers of each other.

> Thus, adjacent bands cannot simply lock-step because a prerequisite for stable temporal locking is phase synchronization. [20]

Neuroscientist György Buzsáki stresses that the brain does not generate pink noise directly. Rather, it produces a wide range of oscillations whose spatio-temporal nesting brings forth pink noise ($1/f$ statistics) [21]. However, according to Buszáki, short-lived states in which the brain exhibits a characteristic temporal scale do occur and also play an important role:

> These transient stabilities of brain dynamics are useful to hold information for some time, as is the case while recognizing a face or dialing a seven-digit telephone number. Shifting the brain state from complex pink-noise dynamics to a state with a characteristic temporal scale is therefore an important mechanism that provides a transient autonomy to various levels of neuronal organization. [22]

Highly periodic behaviour on characteristic scales (i.e. nonfractal) is usually a manifestation of a dynamic disease or disorder. This happens, for instance, in epileptic seizures. In terms of my Theory of Fractal Time, the emergence of mode-locking correlates with the transition of Δt_{depth} into Δt_{length}.

So while the transition from Δt_{length} into Δt_{depth} increases interfacial complexity and is conducive to health, turning Δt_{depth} into Δt_{length} results in a loss of complexity and reduces the range of possible responses. Along with this loss of complexity comes a shift from pink to Brownian noise, which is often due to advanced age. This shift also signifies a reduction in global strong anticipation (which would occur if there were a reduction of nested simultaneity (Δt_{depth}) in the environment, within the human body or on both sides of the interfacial cut.)

Figure 11.3 depicts the departures from complexity due to advanced age. Van Orden *et al.* report that posture and gait diverge towards white noise (bottom of Fig. 11.3) while neural activity (resting fMRI), heartbeat and body temperature deviate towards brown noise (top of Fig. 11.3). A deviation towards white noise has been observed in the heartbeat in atrial fibrillation and gait in Huntington's disease [23].

Morbus Parkinson, a neurodegenerative disorder, is a typical example of multiple deviation towards Brownian noise which shows itself in an increasing rigidity and a loss of complexity in gait and speech [25].

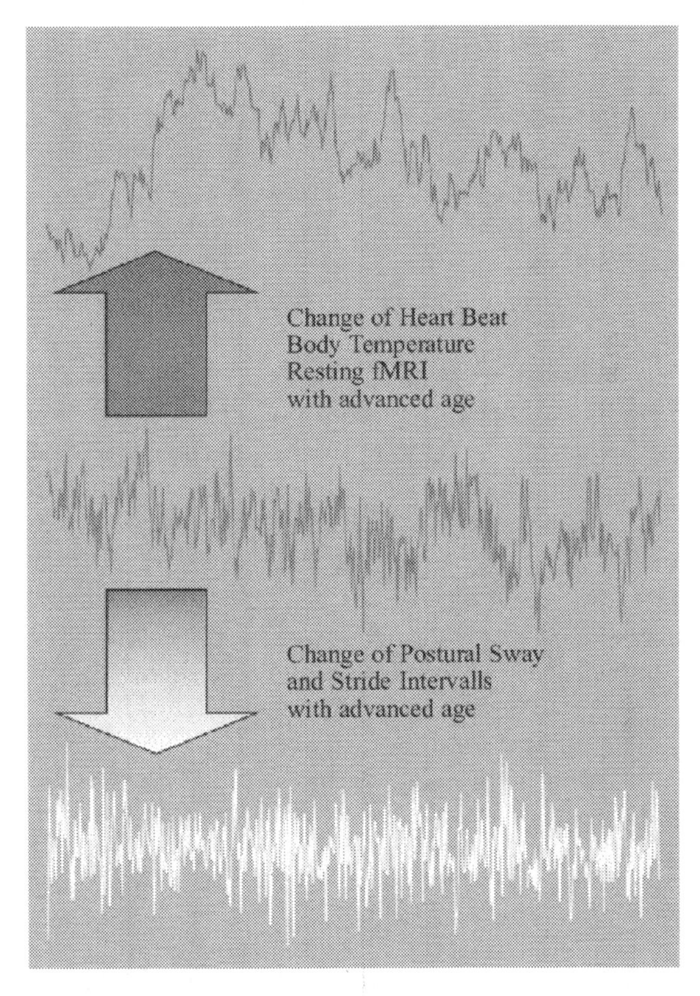

Fig. 11.3. Departures from complexity due to advanced age. The top graph represents Brownian noise, the middle one pink noise and the bottom graph white noise [24].

11.4 Two Types of Local Strong Anticipation

There are two manifestations of local strong anticipation: The first type is obserpant extensions which form a new systemic whole with the incorporated components of their environment. This increases nested simultaneity, i.e. Δt_{depth} (see Chapter 3), and manifests itself as local strong anticipation in the form of delay and distance compensation. It is the compensatory acts which create nested simultaneity, i.e. Δt_{depth}.

Local strong anticipation through compensation of delay and distance is of advantage to obserpants if it widens their sphere of influence. Examples are the transformation of extrapersonal space into peripersonal space and transsensus (see Chapters 3 and 7). In general, compensated delays enable us to navigate the world without constantly colliding with objects or misjudging our fellow human beings' behaviour. However, there are also compensatory acts which are detrimental to health and result, for instance, in the loss of the feeling of body ownership (although a sense of agency may prevail).

The second type of local strong anticipation is anticipatory synchronization, in which coupled oscillators synchronize when an enslaved rhythm is entrained by the rhythm of the master system. As a result, Δt_{depth} decreases because the rhythms are congruent, not nested. Although during the process of synchronization, nested structures occur, the end result, i.e. synchrony, no longer exhibits Δt_{depth}. Such local strong anticipation can be both conducive and detrimental to health. Stephen Strogatz describes coupled oscillators, pointing out their dangers and merits. Healthy, even vital, examples of such coupling abound in nature:

> Groups of fireflies, planets, or pacemaker cells are all collections of oscillators — entities that cycle automatically, that repeat themselves over and over again at more or less regular time intervals. Fireflies flash, planets orbit, pacemaker cells fire. Two or more oscillators are said to be coupled if some physical or chemical process allows them to influence one another. Fireflies communicate with light, planets tug on one another with gravity. Heart cells pass electrical currents back and forth. As these examples suggest, nature uses every available channel to allow its oscillators to talk to one another. And the result of those conversations is often synchrony, in which all the oscillators begin to move as one. [26]

Strogatz stresses that most real oscillators are coupled locally rather than globally. The stomach and intestines are coupled locally along nerve and muscle cells which interact rhythmically. The pacemaker cells in the heart send out electrical impulses to trigger a pulse. These are both examples of useful, even vital, coupled oscillators [27]. But they are also inflexible, which is the reason that pathological dynamics result if individual components generate and maintain their own waves.

Dangerous coupling manifests itself, for instance, in epileptic seizures, convulsions which are caused by millions of brain cells discharging in lockstep [28]. In photosensitive epilepsy, such seizures appear to be caused by brain waves entraining with flickering light. Other pathological coupled oscillators are the pathologic periodicity which governs tremor in Morbus Parkinson and sudden cardiac death [29]. The contagious characteristic of coupled oscillators can lead to collective hallucinations or the dancing plague which hit Europe between 1200 and 1600, causing people to fall into a trance and dance for days or even weeks until they collapsed [30].

These couplings can distort our temporal interfaces by reducing Δt_{depth}. The congruence of simultaneous perceptions leads to fewer or no nesting performances and a decreased level of complexity with a lower fractal dimension ($\Delta t_{\text{density}}$): If all rhythms pulsate in unison, complexity is lost and, with it, the range of possible responses. In other words, a reduction of Δt_{depth} makes us less flexible. Thus, local strong anticipation in the form of anticipatory synchronization manifests itself as a reduction in or even a loss of nested simultaneity, as it creates only congruence. Possible implications of such congruence are dealt with in Chapter 12, where the converse concept, sheer simultaneity, is defined as $\Delta t_{\text{depth}} \to \infty$.

11.5 Recalibration of Δt_{depth} Is Compromised for Obserpants whose Expectations Are Always Met

On the cognitive level, the formation of Δt_{depth} is compromised for obserpants whose temporal interface, i.e. their Now, exhibits a mismatch between retensions and protensions (i.e. memory and anticipation). If there is a bias towards retension, the Now re-contextualizes present perceptions within past ones, rather than integrating them into anticipated ones. Such a retention-heavy Δt_{depth} contextualizes only or primarily in retensions, i.e. an obserpant's nested memories.

If our ability to contextualize in protensions is compromised by a neurodegenerative disease or a mental disorder, our temporal interface's structure becomes less balanced and less complex. It becomes biased towards the past. Psychiatrist and philosopher Hinderk M. Emrich and psychiatrist Detlef E. Dietrich showed that individuals who suffer from depression often perceive the world through a temporal interface which is dominated by the past [31]. They conducted an experiment in which depressed and nondepressed individuals were shown words with positive, neutral and negative emotional content on a screen. Both groups had to rate the words as either old (presented before) or new (presented for the first time), while the corresponding event-related brain potentials (ERP) were recorded. (ERP for expected (old) words were known to be more pronounced.) While the (nondepressed) control group's ERP showed significant differences in the first and second presentation (for both negative and positive words), such differences were hardly detectable in the depressed individuals. Their ERP did not distinguish between old and new. All present stimuli, including new ones, were re-nested as anticipated.

Emrich and Dietrich also showed that depressed individuals were unable to embed negative items in a new (positive) context. This transforms the Now into a temporal structure in which even new stimuli are registered as anticipated (old). They concluded that the cognitive memory system of depressed individuals was primed predominantly by negative cognitions and memories. Depressed individuals anticipate negative stimuli and nest them in an ever-deepening cascade of retension-heavy Nows. This mechanism is a conditioning effect, which increases the negative anticipation in depressed individuals. They lack the ability to perform the positive contextualization necessary to recalibrate the structure of their Now. And if their expectations are always met, any kind of development is stifled. An obserpant's anticipatory faculty is compromised if he or she cannot detect the difference between unexpected (new) stimuli and anticipated (old) ones. (Note that the interpretation of this experiment in terms of extended temporal Nows is mine and does not necessarily reflect the researchers' view.)

11.6 Global and Local Perspectives

If expectations are always met, sad moods tend to resist contextualization in a new, positive, temporal embedding. By contrast, positive moods correlate with contextualization, i.e. nesting performances which are inherent in global perspectives. Psychologists Karen Gasper and Gerald Clore showed in experiments that positive moods generate global perspectives, whereas negative moods give rise to local ones [32]. Participants were shown geometric figures arranged in groups and asked to state whether a target object was more similar to a geometric figure which matched its global arrangement or its local aspects. It turned out that participants who were primed into sad moods tended to see more similarity between the target object and a geometric figure which matched its local structure. By contrast, participants in a happy mood found the target to be more similar to a geometric figure which matched its global structure. The researchers concluded that individuals in happy moods tend to see the forest, while those in sad moods saw the trees.

In terms of my Theory of Fractal Time, happy moods and global perspectives are generated through contextualization in fractal interfaces, thus giving rise to Δt_{depth}. Sad moods and local perspectives, on the other hand, contextualize less and therefore seem to compromise the generation of Δt_{depth} while Δt_{length} increases. When Δt_{length} dominates our interfaces at the expense of Δt_{depth}, our anticipatory faculty is compromised (see Chapter 9). By contrast, an increase in Δt_{depth}, i.e. nested simultaneity, increases our anticipatory faculty. From the correlations in and with Gasper's and Clore's experiments, may we deduce that happy moods create Δt_{depth}, whereas sad moods give rise to Δt_{length}? Further research may shed more light on how moods structure our interface by biasing it towards local or global perspectives.

Global perspectives arise because we contextualize, i.e. we embed structures in a larger framework and thus generate Δt_{depth}. But it is not only the number of nestings which determines our mood and perspective. Nesting speed may be the underlying mechanism which correlates both mood and perspective with the generation of Δt_{depth}. Psychologists Emily Pronin and Daniel Wegner showed that nested thought speed influences moods, disregarding the positive or negative content of the thought [33]. Their experiment had participants quickly reading aloud words which appeared letter by letter and integrating short phonetic units into a meaningful sentence. The cognitive task thus consisted of contextualizing, i.e. nesting syllables into words, words into sentences and sentences into larger contexts. It turned out that the faster the reading, i.e. the faster the rate of contextualization, the more positive the mood of the participants. The correlation of fast thinking with happy moods is reminiscent of the notion that time flies when one is happy and also of manic episodes.

These fast nesting performances generated Δt_{depth} at a faster rate (than those of slow readers), which resulted in a more complex and thus more global perspective and increased anticipatory faculty. Pronin and Wegner suggest that thinking fast may boost our moods even when we have depressing thoughts. This might imply that fast thinking creates happy moods and a global perspective which generates Δt_{depth} and enhances one's anticipatory faculty. If sad moods resist contextualization, i.e. if the formation of Δt_{depth} is compromised by negative moods, they may also constrain our anticipatory faculty.

11.7 Confounded Nesting Arrangement in Δt_{depth} in Schizophrenia

The idea that schizophrenic interfaces differ in the way they contextualize or decontextualize is not new. We saw in Dakin *et al.*'s experiment (see Chapter 9) that schizophrenic individuals failed to perceive simultaneous contrasts, whereas healthy subjects were distracted by them. It helped schizophrenics to better focus on the task at hand (e.g. matching an embedded disk with an isolated one). The difficulty of deciding whose interfacial structure is "correct" becomes apparent in this experiment. The degree to which we are able or willing to contextualize determines the spatio-temporal distortions our interface creates.

Eagleman draws our attention to the fact that many disorders may result from temporal distortions which confound causal order:

> We have recently discovered that a deficit in temporal order judgments may underlie some of the hallmark symptoms of schizophrenia, such as misattributions of credit ("My hand moved, but I didn't move it") and auditory hallucinations, which may be an order reversal of the generation and hearing of normal internal monolog. (...) As the study of time in the brain moves forward, it will likely uncover many contact points with clinical neurology. At present, most imaginable disorders of time

> would be lumped into a classification of dementia or disorientation, catch-all diagnoses that miss the important clinical details we hope to discern in coming years. [34]

Psychiatrist Dieter De Grave points out that schizophrenics do not lack embedding capacity. They have simply created nestings which are incompatible with those created by the majority of obserpants. He defines schizophrenia as "a biopsychological order in its own right" [35]. Their contextualizations follow a closed logic, which is only challenged when internal and external expectations are not met [36]:

> When dealing with schizophrenia, we should always remember that we are talking about people in interaction with their environment, trying to make the best of things. (…) They are busy trying to piece together the missing parts, the things other people tell them are out of order, unhinged, unbecoming in that place, at that time. To the outside world it seems as if time and logic have collapsed into nonsense, to them all forms of understanding have imploded into the magnificent insight which makes their [the schizophrenics'] delusion. (…) It is not that their delusions are wrong *per se*, the fact is that they could be wrong, just like any other thought anybody might have. In this we are all equal and in this we can find a middle ground between 'normality' and schizophrenia. [37]

The debilitating mismatch between internal and external dynamics shows itself in the deficits in schizophrenics' ability to synchronize their actions with external events. Psychologist Hélène Wilquin *et al.* showed this in a tapping experiment and concluded that the reason why schizophrenics fail to synchronize with the outside world is a deficit in predictive timing. And this specific deficit, the authors suggest, may underlie early symptoms of schizophrenia [38].

Impairment of predictive timing may also be at the core of schizophrenics' aberrant sense of agency. This sense is the feeling that one is in control of one's own actions and their effects in the external world. As we saw in Chapter 3, a sense of agency is a necessary condition for local strong anticipation. However, as Hiroki Oi *et al.* pointed out, a lack of anticipation is not the only candidate to account for schizophrenics' symptoms. Impairment in postdictive processes may also contribute to the explanation [39].

Further deficits in anticipatory behaviour have been identified in the lack of anticipatory pleasure (both in schizophrenia spectrum disorders and major depression). Patients' ability to mentally simulate future experiences is compromised, which also discourages their motivation for rewarding behaviour. As this includes psychosocial functioning, improving patients' anticipatory faculty in therapy would greatly enhance their well-being [40].

11.8 Festination in Morbus Parkinson: A Scale-Relativistic View of Global Strong Anticipation

A highly original perspective on the temporal distortions which accompany Morbus Parkinson identifies a mismatch between internal and external spacetime as the culprit of this symptom. Astrophysicist Laurent Nottale's concept of scale relativity (a property intrinsic to the geometry of fractal spacetime) is an extension of the principle of relativity to transformations of scale in the reference system and the measuring device [41]. The scale-relativistic perspective implies that scales of length and time attributed to objects are arbitrary and only the ratio between internal and external scales is a meaningful unit. Consequently, both time and space dilations and contractions are manifestations of scale relativity, which means that, for example, a dilation of the object and a contraction of the unit would be indistinguishable.

Together with psychiatrist Pierre Timar, Nottale suggests a scale-relativistic approach to describing certain manifestations of Morbus Parkinson. Festination is a symptom which shows itself in a shortening of stride that is accompanied by the urge to make much quicker and shorter steps, thus speeding up in a tripping gait. Patients' handwriting becomes contracted, with the letter size decreasing in every line. In speech, words are uttered at a faster rate until it becomes difficult for the listener to make out the individual words. Nottale and Timar suggest that this dilation and contraction results from a "loss of the internal device that synchronizes us to external space-time" [42]. Lucas *et al.* suggest that such changes in time perception result from a deficit in sensorimotor integration in the basal ganglia [43].

Nevertheless, for patients suffering from Morbus Parkinson, festination is an adequate response to their deformed perceptions of spacetime. External pacemakers can help recalibrate internal spatial and temporal spacings. As we saw in Chapter 5, human pacemakers are more efficient than isochronous ones [44]. By synchronizing with an external human pacemaker, individuals increase their range of possible responses. Biostatistician Zainy Almurad *et al.* showed that aged individuals walking in synchrony with young healthy walkers restores complexity in the older one. In most cases, multi-fractality is transferred from the more complex system to the less complex one [45].

This transfer increases the fractal exponents when the old walker internalizes the pacemaker's perception of spacetime. His or her spatio-temporal interface is recalibrated to match external fractal dimensions as expressed in the distributions of succession and simultaneity. And as Morbus Parkinson correlates with Brownian noise, which manifests itself in over-rigid control and a loss of flexibility, restoring the range of responses towards that of pink noise through an external pacemaker would also imply restoring the patient's faculty for global strong anticipation.

To conclude, white noise indicates a loss of structure, whereas Brownian noise exhibits all-too regular dynamics, and both result in a loss of global strong anticipation which is detrimental to health. This is true for heart and gait rate variability as indicators of cardiovascular and neurodegenerative diseases and also for mental disorders. Pathological temporal perspectives are often indicated by psychosis and mood disorders.

In general, global strong anticipation appears to be conducive to physical and mental health. With a few notable exceptions, such as the feeling of a loss of body ownership, local strong anticipation resulting from obserpant extensions is also generally a healthy faculty. By contrast, local synchronization through anticipatory synchronization or direct coupling can be both conducive and detrimental to health: Cardiac pacemakers perform a vital regulative function and human pacemakers improve the gait of patients suffering from Morbus Parkinson, whereas epileptic seizures induced by flickering light have a debilitating effect.

In general, measures which increase obserpants' range of internal responses, i.e. increase the fractal dimension of their spatio-temporal interface, are conducive to health and improve their anticipative faculty.

References

[1] K. Vonnegut, *Slaughterhouse Five*, Dell Publishing, New York, 1991 (first published in 1966), p. 104.

[2] S. Vrobel, How to make nature blush: On the construction of a fractal temporal interface. In: D. S. Broomhead, E. A. Luchinskaya, P. V. E. McClintock and T. Mullin (eds.) *Stochastics and Chaotic Dynamics in the Lakes: STOCHAOS*, AIP, New York, 2000, pp. 557–561.

[3] G. van Orden, H. Kloos und S. Wallot, Living in the pink: Intentionality, wellbeing, and complexity. In: C. Hooker (ed.) *Philosophy of Complex Systems, Handbook of the Philosophy of Science*, Elsevier, Amsterdam, 2011, p. 629.

[4] *ibid*, p. 642.

[5] G. van Orden, G. Hollis and S. Wallot, The blue-collar brain. *Frontiers in Physiology. Hypothesis and Theory*, 3, 6, 18 June 2012. Article 207.

[6] *ibid.*

[7] Illustration and caption: G. van Orden, G. Hollis and S. Wallot, The blue-collar brain. *Frontiers in Physiology. Hypothesis and Theory*, 3, 6, 18 June 2012. Article 207. CC BY 4.0.

[8] G. van Orden, Voluntary performance. *Medicina* (Kaunas), 46(9), 582, 2010.

[9] V. Marmelat and D. Delignières, Strong anticipation: Complexity matching in inter-personal coordination. *Experimental Brain Research*, 222, 137–148, 2012; D. G. Stephen and J. A. Dixon, Multifractal cascade dynamics modulate scaling in synchronization behaviours. *Chaos, Solitons & Fractals* (Elsevier), 44(1–3), 160–168, 2011; G. van Orden, H. Kloos und S. Wallot, Living in the pink: Intentionality, wellbeing, and complexity. In: C. Hooker (ed.) *Philosophy of Complex Systems, Handbook of the Philosophy of Science*, Elsevier, Amsterdam, 2011, pp. 629–672; B. J. West, *Where Medicine Went Wrong — Rediscovering the Path to Complexity,* World Scientific, Singapore, 2006, pp. 283–284; See also Chapter 9.

[10] D. G. Stephen and J. A. Dixon, Multifractal cascade dynamics modulate scaling in synchronization behaviours. *Chaos, Solitons & Fractals* (Elsevier), 44(1–3), 160–168, 2011.

[11] Illustration and caption: G. van Orden, Voluntary performance. *Medicina* (Kaunas), 46(9), 582, 2010. CC BY 4.0.

[12] E.g.: G. van Orden, H. Kloos und S. Wallot, Living in the pink: Intentionality, wellbeing, and complexity. In: C. Hooker (ed.) *Philosophy of Complex Systems, Handbook of the Philosophy of Science*, Elsevier, Amsterdam, 2011, pp. 629–672; D. G. Stephen and J. A. Dixon, Multifractal cascade dynamics modulate scaling in synchronization behaviours. *Chaos, Solitons & Fractals* (Elsevier), 44(1–3), 160–168, 2011; V. Marmelat and D. Delignières, Strong anticipation: Complexity matching in interpersonal coordination. *Experimental Brain Research*, 222, 137–148, 2012; A. Washburn, R. W. Kallen, C. A. Coey, K. Shockley and M. J. Richardson, Interpersonal anticipatory synchronization: The facilitating role of short visual-motor feedback delays. In: *Proceedings of the 37th Annual Meeting of the Cognitive Science Society*, 2015, p. 2619; P. Grigolini, G. Aquino, M. Bologna, M. Lukovic and B. J. West, A theory of 1/f noise in human cognition. *Physica A*, 388, 4192–4204, 2009; B. J. West, *Where Medicine Went Wrong: Rediscovering the Road to Complexity*, World Scientific, Singapore, 2006; Y. Joye, A tentative argument for the inclusion of nature-based forms in architecture. PhD thesis, University of Ghent, Belgium, 2007.

[13] A. Piotrowski, V. Nordmeier and H.-J. Schlichting, Musikalisches Rauschen. In: Bruhn, J. (ed.) *Didaktik der Physik. Vorträge der Frühjahrstagung*, Hamburg, 1994, pp. 355–360.

[14] S. K. Rankin, P. W. Fink and E. W. Large, Fractal structure enables temporal prediction in music. *Journal of the Acoustical Society of America*, 136, EL 257, 2014.

[15] H. Lai *et al.* A new product development and effect analysis of tinnitus therapy based on pink noise tone. *Preprint from Research Square,* January 2021; M. Yamagata *et al.* Subthreshold electrical stimulation with pink noise enhances feedback control as evaluated by scaling exponent of postural sway. *Neuroscience Letters*, 799(6603), 137102, February 2023; A. Golrou *et al.* Enhancement of sleep quality and stability using acoustic stimulation during slow wave sleep. *Clinical Science*, 5(4), January 2019; E. M. Sokhadze, Effects of music on the recovery of autonomic and electrocortical activity after stress induced by aversive stimuli. *Applied Psychophysiology and Biofeedback*, 32, 2007; S. Nair *et al.* Infra-slow pink noise stimulation can increase default-mode network activity in individuals with early Alzheimer's disease. *Conference: 260th Otago Medical School*

Research Society Masters/Honours Speakers Awards at Dunedin, Otago, New Zealand, December 2021.

[16] J. B. Waugh *et al.* Stethoscope transmission characteristics using a pure tone sweep versus pink noise input. In: *Chest* 126 (4_MeetingAbstracts), October 2004.

[17] O. van Nieuwenhuijze, The equation of health. D. M. Dubois (ed.) *International Journal of Computing Anticipatory Systems* (CHAOS, Liège), 22, SS. 235–249, 2008.

[18] B. J. West, *Where Medicine Went Wrong — Rediscovering the Path to Complexity*, World Scientific, Singapore, 2006, pp. 283–284; A. L. Goldberger, L. A. N. Amaral, J. M. Hausdorff, P. C. Ivanov and C.-K. Peng, Fractal dynamics in physiology: Alterations with disease and aging. *PNAS Colloquium*, 99(Suppl. 1), 2466–2472, 19 February 2002.

[19] A. L. Goldberger, L. A. N. Amaral, J. M. Hausdorff, P. C. Ivanov and C.-K. Peng, Fractal dynamics in physiology: Alterations with disease and aging. *PNAS Colloquium*, 99(Suppl. 1), 2471, 19 February 2002.

[20] G. Buzsáki, *Rhythms of the Brain*, Oxford University Press, New York, 2006, p. 120.

[21] *ibid*, p. 130.

[22] *ibid*, p. 131.

[23] *ibid*, pp. 283–284; Hausdorff *et al.*, Altered fractal dynamics of gait. Reduced stride-interval correlations with aging and Huntington's disease. *Journal of Applied Physiology*, 82, 262–269, 1997; G. van Orden, H. Kloos und S. Wallot, Living in the pink: Intentionality, wellbeing, and complexity. In: C. Hooker (eds.) *Philosophy of Complex Systems, Handbook of the Philosophy of Science*, Elsevier, Amsterdam, 2011, p. 662.

[24] G. van Orden, H. Kloos und S. Wallot, Living in the pink: Intentionality, wellbeing, and complexity. In: C. Hooker (eds.) *Philosophy of Complex Systems, Handbook of the Philosophy of Science*, Elsevier, Amsterdam, 2011, p. 662, CC BY 4.0.

[25] *ibid*, p. 662.

[26] S. Strogatz, *Sync — How order emerges from chaos in the universe, nature and daily life*, Hyperion, New York, 2003, p. 3.

[27] *ibid*, p. 210.

[28] *ibid*, p. 14.

[29] A. L. Goldberger, L. A. N. Amaral, J. M. Hausdorff, P. C. Ivanov and C.-K. Peng, Fractal dynamics in physiology: Alterations with disease and aging. *PNAS Colloquium*, 99(Suppl. 1), 2466–2472, 19 February 2002.

[30] J. Waller, *A Time to Dance, a Time to Die. The Extraordinary Story of the Dancing Plague of 1518*, Icon Books, London, 2008.

[31] H. M. Emrich and D. E. Dietrich, On time experience in depression — dominance of the past. In: T. Schramme and J. Thome (eds.) *Philosophy and Psychiatry*, 2005, pp. 242–256; H. M. Emrich, C. Bonnemann and D. E. Dietrich, On time experience in depression. In: S. Vrobel, O. E. Rössler and T. Marks-Tarlow (eds.) *Simultaneity — Temporal Structures and Observer Perspectives*, World Scientific, Singapore, 2008.

[32] K. Gasper and G. L. Clore, Attending to the big picture: Mood and global versus local processing of visual information. *Psychological Science*, 13(1), 34–40, 2002.

[33] E. Pronin and D. M. Wegner, Manic thinking. Independent effects on thought speed and thought content on mood. *Psychological Science*, 17(9), 807–813, 2006.

[34] D. M. Eagleman, *Brain Time*, 2023 by Edge Foundation. Accessed 2.4.23. https:// www.edge.org/conversation/david_m_eagleman-brain-time.

[35] D. De Grave, The readiness is all? Closure remarks on the psychotic anticipatory experience of time and space. Doctoral Dissertation, Ghent, Belgium, 2013, pp. 217ff.

[36] D. De Grave, personal communication, 2005.

[37] D. De Grave, The implosion of reality. Schizophrenia, the anterior cingulate cortex and anticipation. D. M. Dubois (ed.) *International Journal of Computing Anticipatory Systems*, 22 (Liège Belgium), 2006, 161, 2008.

[38] H. Wilquin, Y. Delevoye-Turrell, M. Dione and A. Giersch, Motor synchronization in patients with schizophrenia: Preserved time representation with abnormalities in predictive timing. Frontiers in Human Neuroscience, 12, 2018. Article 193.

[39] H. Oi, W. Wen, M. Mimura and T. Maeda, Evaluating weightings of predictive processes in aberrant sense of agency in schizophrenia. *Schizophrenia Bulletin*, 46(Suppl. 1), 156–157, May 2020.

[40] D. J. Hallford and M. K. Sharma, Anticipatory pleasure for future experiences in schizophrenia spectrum disorders and major depression: A systematic review and meta-analysis. *British Journal of Clinical Psychology*, 2019.

[41] L. Nottale, Relativité d'échelle structure de la théorie. *Revue de synthèse*, 122, 11–25, 2001; L. Nottale, Scale relativity, fractal space-time and morphogenesis of structures. In: H. H. Diebner, T. Druckrey and P. Weibel (eds.) *Sciences of the Interface*, Genista, Tübingen, 2001, pp. 38–51.

[42] L. Nottale and P. Timar, Relativity of scales: Application to an endo-perspective of temporal structures. In: S. Vrobel, O. E. Rössler and T. Marks-Tarlow (eds.) *Simultaneity — Temporal Structures and Observer Perspectives,* World Scientific, Singapore, 2008, p. 235.
[43] M. Lucas, F. Chaves, S. Teixeira, D. Carvalho, C. Peressutti, J. Bittencourt, B. Velasques, M. Menéndez-González, M. Cagy, R. Piedade, A. E. Nardi, S. Machado, P. Ribeiro and O. Arias-Carrión: Time perception impairs sensory-motor integration in Parkinson's disease. *International Archives of Medicine*, 6, 39, 2013.
[44] V. Marmelat, D. Delignières, K. Torre, P. J. Beek and A. Daffertshofer, 'Human paced' walking: Followers adopt stride time dynamics of leaders. *Neuroscience Letters* (Elsevier), 564, 2014.
[45] Z. M. H. Almurad, C. Roume, H. Blain and D. Delignières, Complexity matching: Restoring the complexity of locomotion in older people through arm-in-arm walking, *Frontiers in Physiology*, 9, 2018. Article 1766.

Chapter 12

Insight and Nonlocality as Strong Anticipation

As I mentioned in earlier chapters, the distinction between temporal and spatial extensions is, of course, not tenable, as we are always dealing with spatio-temporal phenomena. Compensating delay and distance in spacetime generates local strong anticipation through the coupling of organism and environment (or two organisms).

Global strong anticipation evolves from the coordination of long-term correlations, as those exemplified in Stephen and Dixon's tapping experiments and the matching of inter-personal coordination described by Marmelat and Delignières (see Section 2.6.2). These long-term nonlocal correlations are of a probabilistic and global nature, as are the results of Korotaev *et al.*'s Baikal experiments (to be described in this chapter). The resulting nonlocal correlations, which Einstein referred to as "spooky action at a distance", are interpreted as global strong anticipation.

Another manifestation of global strong anticipation is the phenomenon of insight, as outlined by philosopher Jiddu Krishnamurti and my notion of a condensation scenario. Sheer simultaneity manifests itself as time condensation in self-similar fractal time series when Δt_{depth} approaches ∞.

Fractal spacetime, as portrayed by Nottale and El Naschie, belongs both to the realms of phenomenology and ontology. On the one hand, the fractal geodesics of spacetime constrain, as an ontological entity, the microscopic movements within us as well as those in our environment. On the other hand, our embodied brains have come up with the concept of fractal spacetime as a result of our co-evolution with a phase of the universe. Phenomenology and ontology are inextricably interwoven in our interfaces because, as obserpants, we simultaneously partake in both the quantum and classical levels with which we co-evolved. My own approach remains phenomenological and systems-theoretic. At the bottom line, insight and nonlocal perspectives may eradicate phenomenal and epistemological blind spots but come at a price — the end of naïve realism and smooth navigation.

12.1 Nonlocal Correlations: Is Spooky Action at a Distance Just a Compensated Delay?

Sir Isaac Newton's correspondence with Doctor Bentley revolved around proof of a deity and the nature of gravity. On 25 February 1692, Newton urged Bentley not to ascribe the notion of innate gravity to him, as the idea of matter affecting other matter without mutual contact was inconceivable to him:

> That Gravity should be innate, inherent and essential to Matter, so that one Body may act upon another at a Distance thro' a Vacuum, without the Mediation of anything else, by and through which their Action and Force may be conveyed from one to another, is to me so great an Absurdity, that I believe no Man who has in Philosophical Matters a competent Faculty of thinking, can ever fall into it. Gravity must be caused by an Agent acting constantly according to certain Laws; but whether this Agent be material or immaterial, I have left to the Consideration of my Readers. [1]

Newton was well aware that his notion of gravity implied instantaneous action at a distance, a problem he could not solve and on which he did not wish to speculate.

Einstein acknowledged that Newton's model works locally but provided a completely new description of how masses attract each other.

In his general theory of relativity, he described gravity as a property of the geometry of spacetime. Geodesics, which locally appear straight (for instance, on our planet), approach each other in larger regions of spacetime. This geometric effect on matter, according to Einstein, is gravity. It shapes the curvature of spacetime locally and does not take effect instantaneously. Theoretical physicist John Wheeler put it in a nutshell:

> Space acts on matter, telling it how to move. In turn, matter reacts back to space, telling it how to curve. [2]

He also pointed out that the world lines which penetrate Minkowski space (a four-dimensional manifold to represent relativistic spacetime, in which each point represents an event) represent different simultaneity horizons in a slab of jello, but

> spacetime does not wiggle. It is 3-D space geometry that undergoes agitation. The history of its wiggling registers itself in frozen form as spacetime. [3]

Quantum mechanics refuted Einstein's concept of a causal and locally acting gravity and elevated nonlocality to one of its hallmark features. Physicist John Stewart Bell discovered that pairwise correlated particles in a sum-spin-zero state were superluminally, i.e. nonlocally, connected so that by measuring one and thus destroying the superposition, the other particle's state was determined instantaneously. Einstein refused to acknowledge the existence of nonlocality and called it "spooky action at a distance". The so-called EPR paradox, a hypothetical scenario described by him, Podolsky and Rosen [4], questioned whether a quantum mechanical description of the world could be complete. As quantum physicist Federico Levi puts it:

> The trio questioned the ability of quantum mechanics to comprehensively link the evident objective reality to suitable theoretical concepts. The whole argument, in modern terms, hinged on the fact that for an entangled pair of particles, selecting an observable to measure at one

> end alters which property should be considered real at the other end — without any interaction necessary. [5]

The counterintuitive concept of reality revealed by the EPR paradox is now widely accepted and the majority of the scientific community has conceded that nonlocality lies at the heart of quantum mechanics. However, as Levi remarks,

> (...) the true legacy of the EPR paradox lies in the investigation of the relationships between locality and reality, which remains far from exhausted. [6]

In defence of Einstein's scepticism of Bell nonlocality, Rössler suggested that for a specific interpretation of Hugh Everett's many-worlds theory, nonlocality would no longer imply superluminal connections [7]. Everett's theory is a local and deterministic interpretation of quantum mechanics, which sees the uncollapsed state vector as an objective description of the world. His theory allows for all possible interfacial cuts and measurement results as many private simultaneous obserpant perspectives. It avoids the wave function collapse which implies that any interaction (measurement/observation) reduces all possible worlds to one [8]. Everett's interpretation was in stark contrast to the Copenhagen interpretation and thereby to the interpretation to which most of the scientific community adheres.

An interesting view was brought forward by Wheeler, who saw the role of the obserpant as that of an active, selecting questioner, whose subjective choice of questions generate reality. He invented a new version of the 20-questions game: Instead of a panel agreeing on a secret word the candidate has to guess with no more than 20 questions, the members of the panel have to make it up as the game proceeds. So, after answering the questions "Is it edible?", "Is it a mineral?", etc. with "yes", the panel members have to think of a word which matches all the selections made so far. The word is brought into existence by the chain of questions and answers. Although the analogy does not quite hold, it hints at Wheeler's idea of the participatory anthropic principle: In the measuring process, the obserpant plays a constituting role in the transition from the possible to the actual. The window in which the decision on the way of measuring and the measurement itself is taken, i.e. our Now, is understood as an indivisible whole consisting of particles and the experimental setup. What has become known as the "delayed-choice experiment" (a variation of the double-slit experiment) showed that a measurement made in the present can determine the past manifestation of a particle (its location or its property). Wheeler concludes that the experimenter's delayed choice of the experimental setup retroactively alters the particle's previous behaviour:

> We are participators in bringing into being not only the near and here but the far away and long ago. [9]

The predictions of Wheeler's delayed-choice *gedanken experiment* have been verified in a number of experimental setups [10]. Wheeler went further than just assuming a nonlocal effect in the delayed-choice experiment. His participatory anthropic principle assigns the creative act to the obserpant:

> We used to think that the world exists 'out there' independent of us, we, the observer safely hidden behind a one-foot thick slab of plate glass, not getting involved, only observing. However, we've concluded in the meantime that that isn't the way the world works. In fact, we have to smash the glass, reach in, install a measuring device. (...) We are inescapably involved in coming to a conclusion about what we think is already there. [11]

It appears as if the delay compensation which is initiated by the experimenter's delayed choice is postdictive rather than anticipatory. But there are actually two possible explanations:

(i) Delayed choice is *de facto* the removal of an existing delay. However, unlike in the second keypress experiment, the delayed choice experimenter has not been conditioned to the delay before he makes a choice. Therefore, no compensatory act is involved and thus it is not a case of strong anticipation.

(ii) Delayed choice is *de facto* the removal of an existing delay. The delayed choice experimenter is not aware of the fact that he or she has already compensated the delay before making a choice. The compensated delay is a phenomenal blind spot and a manifestation of strong anticipation.

If (ii) applies, it remains a matter of speculation whether this transparent compensation is a selection effect which favours strong anticipation in obserpants. So does the assignment of the origin of compensated delays to David Bohm's notion of an implicate order, i.e. the totality of higher contexts from which the universe unfolds [12]. For our purposes, the above distinction shall suffice for the reader to decide whether the delayed-choice experiment is a case of postdiction or anticipation.

Most interpretations of quantum theory assign a reality-constituting role to the obserpant but regard the obserpant's microscopic internal differentiation as irrelevant to the experimental setup. By contrast, Rössler's interpretation of the many-worlds theory follows Bell's proposed version of Everett's theory in which the observer is surfing across different Everett worlds [13]. Unlike in Everett's original versions, however, the universes would not exist simultaneously in an unphysical dimension but successively along an ordinary time axis. This rapid change of worlds would go unnoticed, as the whole world, including the obserpant would have changed, leaving no trace of the preceding worlds. Rössler suggests that nonlocality is brought about by a property of the obserpant. In this interpretation, Rössler derives quantum mechanics from a microscopic relative state [14]. Micro-relativity assumes that the interfacial cut between obserpant (as a dissipative system) and the rest of the world affects spacetime physics on a microscopic level. Every microscopic motion within the obserpant affects the corresponding interface-specific worlds:

> … the quantum world is tailor-made for the observer since it only exists as the 'difference' between observer-state and rest-of-the-world-state (interface reality). [15]

Nonlocality would thus be a property of the obserpant's interface. To avoid solipsism, Rössler proposes an endophysical interpretation based on an interfaciological approach. He concludes that the individual quantum worlds are "interface-bound" manifestations of a single exo-reality. Therefore, manipulating the obserpant–world interface brings about a change in the world. But if the entire world including the obserpant changes, no information from the preceding worlds would be accessible. So even if a world-changing interface technology were conceivable, we would not notice the effects. Our Now is the interfacial cut, and despite the world-hopping effects, our present temporal perspective would always appear consistent [16]. This consistency could, of course, also be an intra-dream consistency — after all, we may all be dreaming. Descartes asked himself how to find out whether he is dreaming or awake and concluded that he would simply have to look for inconsistencies. An inconsistent world would reveal the dream and point to the existence of an exteriority in the form of an awake state [17]. And if one judged nonlocality to be inconsistent with the prevailing paradigms, its existence would hint at the presence of an exteriority.

How could we find out whether microscopic nonlocality is interface-induced? Rössler endorses Legget's idea that a quantum Foucault pendulum could prove the existence of some externality: the world as it would appear without interfacial cuts [18]. He conceded that we would still be barred from Kant's *Ding an sich*, i.e. seeing the world objectively from the perspective of an external obserpant. However, by manipulating the "second causation", the micro-interface, we would be able to construct a world-changing technology. (In addition to natural laws and initial conditions as the first types of causation, Rössler defined assignment conditions as a second causation. This second causation is our micro-interface, which can be manipulated, for instance, by changing the difference in internal and external temperatures [19].)

Another side of the constituting role of the obserpant was proposed by Mihai Nadin with his "semiotic engine" and the relation between nonlocality and anticipation:

> Let me [...ascertain ...] that anticipation is a particular form of non-locality, which is quite different from saying that there is non-locality in anticipation. (This is what actually distinguishes my thesis from the results of Dubois.) More precisely, its object is co-relations (over space

> and time) resulting from entanglements characteristic of the living, and eventually extending beyond the living, as in the quantum universe. These co-relations correspond to the integral character of the world, moreover, of the universe. Our descriptions ascertain this character and are ultimately an active constituent of the universe. We introduce in this statement a semiotic notion of special significance to the quantum realm: Sign systems not only represent, but also constitute our universe. [20]

The nontemporal character of sign systems and the temporal constituting interpretant are the basis of Nadin's "semiotic engine" (a computer with iconic, indexical and symbolic descriptions of itself and the world). He emphasizes that anticipatory computation assumes that "every sign is in anticipation of its interpretation" and sign systems are constructed and interpreted by the intelligent observer of quantum mechanics. For his semiotic engine, Nadin rewords the arrow from object to representamen to interpretant in Peirce's sign definition: "The interpretant as a sign refers to something else anticipated in and through the sign" [21].

If Rössler is correct and microscopic nonlocality were interface-induced, what about macroscopic nonlocal correlations? In the world of physics, microscopic nonlocality has been shown to exist not only between entangled particles with spatial separation but also for temporally separated ones. Quantum engineer Eli Megidish *et al.* entangled two photon pairs which were temporally separated, i.e. never coexisted, through entanglement swapping [22]. They emphasize that their results may be interpreted as the present altering either the past or the future:

> In the standard entanglement case, the measurement of any one of the particles instantaneously changes the physical description of the other. This result was described by Einstein as 'spooky action at a distance'. In the scenario present here, measuring the last photon affects the physical description of the first photon in the past, before it has even been measured. Thus, the 'spooky action' is steering the system's past. Another point of view that one can take is that the measurement of the first photon is immediately steering the future physical description of the last photon. In this case, the action is on the future of a part of the system that has not yet been created. [23]

If spatially and temporally separated entities are defined as components of systemic wholes, bridging that separation is an example of compensating a delay, i.e. of strong anticipation. Whether the entanglement was intentionally created or already existed through a preordained order is irrelevant to the compensatory act.

Microscopic nonlocal effects have been experimentally demonstrated, for instance, in teleportation experiments [24]. Macroscopic nonlocality, by contrast, is a different story altogether.

Physicist Sergey Korotaev *et al.* conducted a series of experiments on macroscopic nonlocality which allow long-term forecasting of random solar activity [25]. Korotaev and others also detected nonlocal correlations in other dissipative processes, such as seismic activity [26].

Korotaev *et al.* conducted their experiments over a period of many years at the base of the Baikal Deep Sea Neutrino Observatory (as of 2023, the project was still under way). The thick water layer of the lake shields local impacts on the detectors, which were located at a depth of 47 and 1337 meters. The electrode-type detectors [27] operated autonomously for a year, then the data were read and batteries exchanged every March for the next year. The researchers found nonlocal correlations between the detector signals and the large-scale dissipative processes in solar activity. The lower detector (the one placed at a depth of 1337 meters) predicted the beginning of a new solar activity cycle with an advancement of 315 days. Although geomagnetic activity is a direct effect of solar processes, the relative delay in geomagnetic activity spans only 1 or 2 days, which is far too short to be significant to the experiment's time scales [28].

The Baikal experiment reveals a manifestation of macroscopic entanglement in the detector's advanced response to dissipative heliogeophysical processes. Other random processes, such as earthquakes, also triggered advanced detector activity. The lower detector registered the Mongolian earthquake of 11.1.2021 with a magnitude of 6.5 in advance, 7.5 hours before the event. Korotaev *et al.* conclude:

> The advanced nonlocal correlation can be used to forecast the random components of these natural processes. Such an implementation of

> nonlocal correlation has been demonstrated in the forecast of the beginning of a new solar activity cycle and on long forecasting series of sea current velocity. The advanced response of the nonlocal correlation detector on a strong earthquake can be used (...) for the short-term forecast of such events too. [29]

The distance and delay compensation of the lower detector in Korotaev *et al.*'s experiments is an example of strong anticipation. In addition to demonstrating the classically forbidden notion of predicting random processes, the Baikal experiments delivered more surprising data:

> The Baikal deep sea experiment has provided and continues to provide multifaceted information on nonlocal correlations of large-scale random dissipative processes. These correlations contain retarded, quasi-synchronous and advanced components, which are manifested in causality, independence and correlation functions. For short-term powerful events they can be visible directly in the detector signal. [30]

How to interpret these data? There is a theoretical, albeit controversial, explanation of retarded and advanced signals and one based on equally controversial experimental results. The explanation based on advanced and retarded signals stems from Cramer's interpretation of the (time-symmetric) Wheeler–Feynman absorber theory [31]. Wheeler and Feynman's theory suggests that the electromagnetic field equations are invariant under time-reversal transformation. They do not exhibit any bias towards the past or the future. In Cramer's nonlocal transactional interpretation of quantum mechanics, both the source and the receiver emit a retarded wave, which propagates forward in time, and an advanced wave in reverse time. (The transaction takes place when advanced and retarded waves generate a quantum "handshake", exchanging their properties such as energy and momentum.) The advanced correlation always exceeds the retarded one, as the absorption efficiency of the two differs. This was also observed in the Baikal experiments.

The experimental explanation is based on results obtained by astrophysicist and astronomer Nikolai Aleksandrovich Kozyrev in the late 1970s in his observation of celestial bodies and galaxies. Together with

V.V. Nasonov, he detected not only advanced nonlocal interstellar correlations. They observed the same nonlocal correlations when they directed the telescope not at the visible but at the true position of a galaxy, the region the galaxy "occupied" in "real time", i.e. without signal transmission time elapsing. Next, the signal also appeared when the telescope was shifted and directed at a position symmetrical to the visible position of a galaxy relative to its true position, i.e. it was directed at a future position of the galaxy. This means that the detector registered the past, present and future positions of the galaxy. The researchers concluded that while it was not possible to predict the future, we may be able to observe it [32].

The Baikal experiments were motivated by Kozyrev's experimental results, although Korotaev *et al.* did not interpret them in semi-classical terms as did Kozyrev in his causal mechanics [33].

But however one wishes to interpret the Baikal experiments' results, the long-term nonlocal correlations discovered by Korotaev *et al.* are of a probabilistic and global nature, as are those described by Stephen and Dixon, Marmelat and Delignières, and others [34]. The coordination of these correlations is an example of global strong anticipation.

The Baikal experiments encourage further research into time-symmetrical long-term nonlocal correlations that exhibit advanced, synchronous and retarded components. For our purposes, the advanced components are of interest as examples of global strong anticipation.

12.2 Insight as Strong Anticipation: Sheer Simultaneity

So far, we have seen examples of strong anticipation in Dubois' concept of delay compensation, anticipatory synchronization, in distance and delay compensation through obserpant extensions and interfacial shifts, in the coordination of internal and external 1/*f* (pink) noise, possibly in Wheeler's delayed choice phenomenon and in Korotaev *et al.*'s advanced correlations.

Another type of strong anticipation is achieved through a phenomenon philosopher Jiddu Krishnamurti denotes and describes as *insight* [35]. In his dialogue with David Bohm, he claims that our tendency to think prevents us from brain-restructuring insight. The brain with its successive thinking is caught in time, whereas the mind can operate in a timeless realm. In order to rid ourselves of time-laden thought, we have to stop accumulating time by focusing our life on "the outward". When we focus on the outward, we simulate it inside ourselves, so the inward movement becomes the same as the outward movement. Krishnamurti claims that only when all inward movement ceases, the inside and outside worlds are no longer separated and we can partake in a universal mind and an eternal Now.

There are numerous ways of filtering out environmental stimuli in order to calm internal dynamics. Hove *et al.* showed that trance induction correlates with perceptual decoupling through dampened sensory processing. Because the brain suppresses predictive sensory input, repeating a mantra in meditation helps decouple from the external world [36]. By creating a predictive, i.e. anticipated, environment, the obserpant can provide fertile grounds for an experience of insight. Similar results can be achieved in isolation tanks, in which obserpants float in a light- and sound-proof container in skin-temperature water, to induce sensory deprivation.

However, strictly speaking, any attempt to fully decouple ourselves from our environment is doomed to fail. It would be extremely laborious to escape the Earth's gravitational pull (only to move into the attractive basin of another inertial system). Even if it were possible to filter out all external stimuli, it would be impossible to drown out the inherent internal noise caused by the microfluctuations within our body. In Rössler's interfaciology, these microscopic movements shape our interface with the rest of the world and determine our perspective of the exo-world. Media artist Peter Weibel called these movements "the noise of the observer" [37]. Internal and external noise is inextricably intertwined and produces the interference pattern we perceive as our world. Weibel pointed out a possible danger if the obserpant interprets his or her own internal noise as information about the external world. In this case, hallucinations would emerge, as occur in sensory deprivation. It seems that Krishnamurti's aim of ceasing all inward movement can be approached but not fully achieved [38].

However, this does not mean that insight cannot be attained. In fact, the internal noise may be necessary to create a state in which all succession ends. After all, to Krishnamurti, timelessness is simply the absence of succession. Elsewhere, I have described a *gedanken experiment* in which insight is attained as time condensation through the generation of sheer simultaneity and the absence of succession [39]. If a nonfractal obserpant locked into an external oscillation by internally simulating this frequency, the resulting coordination between the inside and outside would show itself as local synchronization limited to one time scale. As the obserpant's interface would not contain any embedding or embedded structures, Δt_{depth} would remain constant. However, if an obserpant with a temporal fractal interface coordinated with a temporal fractal environment, correlations and synchronization would occur on many nested levels. An example of such multi-level coordination is Erimaki *et al.*'s observation that obserpants exposed to a fractal structure such as the Mandelbrot set exhibit increased synchronization [40]. The obserpant's fractal interface meets a fractal environment. Erimaki *et al.* suggest that this increased synchronization may hint at the existence of a neurobiological basis for human beings' preference for fractal structures, as opposed to nonfractal ones [41].

But how can such multi-layered synchronization between the inside and outside trigger the experience of insight? Descriptions of insight often imply instantaneousness, unexpected sudden flashes which appear to happen in a Now which is extended but has no duration. This "Eureka moment" of sudden recognition of cause and effect or sudden solutions to complex questions can result from extensive training. Science journalist Malcolm Gladwell based the phenomenon of insight and gut decisions which happen within the blink of an eye on the effect of the adaptive unconscious [42]. However, attempts to explain the immediate and embodied form of understanding often remain elusive. An example which I would tentatively assign to the blink effect is physicist Roger Penrose's description of his experience of insight, which he insists was instantaneous, nonalgorithmic and nontemporal. While crossing a street and pondering an unrelated topic, the solution to a problem about the point of no return during the collapse of black holes revealed itself to him in a flash [43].

Such instantaneous, compact understanding can be modelled with the temporal dimensions of my Theory of Fractal Time. If no duration is involved in insight, no succession in terms of Δt_{length} is generated.

The instantaneous, compact density of the content which the obserpant absorbs in a flash corresponds to a high density generated by simultaneous nestings, i.e. to an extensive Δt_{depth}. Thus, insight can be expressed in terms of the temporal dimension succession and simultaneity as $\Delta t_{\text{depth}} \rightarrow \infty$ and $\Delta t_{\text{length}} \rightarrow 0$. Elsewhere, I have defined this relation between Δt_{depth} and Δt_{length} as a condensation scenario [44]. What would remain on the innermost level of a fractal structure is a temporal natural constraint, which I have denoted as the Prime. Within the range of scaling structures, the Prime is the smallest interval in a temporal nesting cascade. It is defined as a nesting constraint, as it cannot host any further nestings. In natural fractal structures, scaling behaviour is usually limited by upper and lower boundaries. Nottale equates such a constraining lower length scale with the Planck scale and the upper limit with the cosmic length scale which is related to the cosmological constant [45].

The indivisible Prime is defined as an interval which cannot host further nestings: It is a nesting constraint. As the structure of the Prime recurs on all levels within a nesting cascade, it can be set as a constant (the Prime Structure Constant (PSC)). In a condensation scenario, the temporal extensions the structure covers on the nested levels are distorted and a scale relativity emerges: Time is "bent" with respect to the PSC.

If the units on all levels of the nesting cascade can be converted into each other, i.e. if a PSC exists, condensation can be quantified as condensation velocity $v(c)$ and condensation acceleration $a(c)$. The quotient of Δt_{length} of the embedded level and Δt_{length} of the embedding level equals the condensation velocity $v(c)$ for those two nested levels. A mathematical scale-invariant structure, such as the Koch curve, $v(c)$ is equal to the scaling factor $s = 3$. In this case, the condensation acceleration would be constant: $a(c) = 1$ [46].

A possible candidate for an insight-inducing condensation scenario is an artificially generated fractal time series so as to match internal nested oscillations with external ones and vice versa. A multi-level one-to-one mapping of interface and environment in terms of nested structures may be experienced as insight in the form of sheer simultaneity, with Δt_{depth} approaching infinity and Δt_{length} ceasing into nothingness, as there would be no more succession. This type of insight would also entail an act of strong anticipation, as succession is turned into simultaneity (see Chapter 5). Instantaneous insight would require congruence on all nested levels. A strictly self-similar interface and environment may produce a local coupling between the nested layers and thus local strong anticipation. However, only ideal, mathematical fractals exhibit exact self-similarity. Natural fractals display statistical self-similarity, such as $1/f$ noise. A matching of internal and external statistical fractal structures (such as pink noise) would generate an interference pattern which prevents phase-locking. Global strong anticipation would result (see Chapters 5 and 11) but without a condensation scenario.

So how to explain Penrose's Eureka moment? In order to end all division between internality and externality, to trigger a nonalgorithmic and nontemporal access to a revealing insight, the Prime would have to be the only structure extending our interface. As Penrose had for a while been pondering the problem of the point of no return during the collapse of black holes from all possible angles, his temporal interface already contained a nested structure of retensions and protensions of this very conundrum. The intersection of contemplations of the problem from different angles would result in something like the Prime Structure Constant, albeit not with mathematical precision. The simultaneous perception of all nested intersections may have allowed him to see a condensed version of a variety of possible approaches to a solution. So, my speculative explanation of Penrose's insight would not entail a complete condensation scenario but a condensation of an intersection which served as a semantic PSC.

Through continuously embedding an impression or a vague notion in a cascade of ever-new contexts in our Now, we generate concepts. When Otto Rössler asked me in 1994 whether I think that "redness" is also a Prime, I had no idea what he meant. I had just defined the concept of the Prime, together with the PSC, in a draft of my doctoral dissertation but did not grasp the extent of his question. It took many decades of circling the thought until I finally thought I saw what he meant: When we first perceive colours, someone tells us that a certain object is red. The next objects we assign the colour red to we do so by trial and error: If someone confirms he or she also assigns the property redness to an object I have pointed out, the notion of redness emerges, with every layer of retension and protension, in a more and more refined way. The constructivist approach already hints at the fact that redness belongs to the realm of qualia and as such has no direct relation to the outside world. The notion of redness results from the continuous nesting of retensions and protensions, which structure our fractal Now. It is the intersection of all nested redness-contexts which constructs my notion of the colour red.

A similar intersection of a theme in various contexts can be seen in paintings, where the artist creates different impressions of a motif, depending on the context he or she has placed it in. Every good portrait is the compact condensation of the motif as it appeared to the artist in various contexts and embedded in his or her retensions and protensions. In the process of creation, Δt_{length}, which was accumulated during the act of observing and painting, has ceased to exist as such in the end product but is still visible in the condensed form of Δt_{depth}'s embeddings.

To conclude, global strong anticipation manifests itself in nonlocal relations. If nonlocality is a property of the obserpant's micro-interface, manipulating its structure would, according to Rössler, result in a world-changing technology. A quantum Foucault pendulum might even reveal the existence of an exteriority — an exo-world which exists independent of the obserpant and which we cannot access directly. Our only access is the interfacial cut through which we participate in it.

Strong anticipation also reveals itself as insight in the cessation of succession through the conversion of Δt_{length} into Δt_{depth}. This may happen in meditative states, when the obserpant focuses on one nesting level, during sensory deprivation or in a condensation scenario [47]. Strong anticipation is of a local character if the matching between interfacial and environmental structures consists of fractals with exact scaling properties. If the matching is of a statistical nature, global strong anticipation can emerge.

The interfacial complexity reduction, which results from the fractal structure of our Now, is custom-made to and has co-evolved with our environment. Multi-layered matching facilitates communication, navigation and science. As we saw in Chapter 11, it is also conducive to health.

12.3 Fractal Time and Fractal Spacetime: Phenomenology or Ontology?

In the introduction to this book, I characterized my approach as phenomenological and systems-theoretical. In this book, I have followed Husserl's insight that phenomenology is the basis of knowledge and philosophy, thus preceding both ontology and epistemology. Our intentions and the structure of our perceptual apparatus determine what we construct as objects of experience. Models based on relations between objects of experience are constructed and interpreted phenomenologically [48].

The fact that researchers from very different walks of life have developed theories of scale-relativity and fractal spacetime begs the question of why our necessarily anthropocentric theories about the world should focus on the resolution of fractal spacetime.

One answer is, of course, that as embodied obserpants with fractal interfaces, whose complexity is determined by the distribution of Δt_{length} and Δt_{depth}, we have co-evolved with our environment. Our interfacial constraints, which have evolved from our experience as embodied agents, now shape our tools, methods, theories and paradigms. And as we have seen in the preceding chapters, both our interface and our environment display fractal patterns. Thus, the fractal obserpant tunes into his or her environment, both on the microscopic and the macroscopic scale.

But are we primed by the fractal geodesics of spacetime so that both pattern formation and pattern recognition result from this preconditioning? When we talk about fractal spacetime, we are dealing with the realm of high-energy physics. El Naschie's concepts of Cantorian spacetime and E-Infinity are based on a fractal spacetime we observe simultaneously at different levels of resolution. It is a logical space in which time is spatialized [49]. However, the temporal component is re-introduced by the obserpant, who creates a private perspective of this fractal spacetime, as each obserpant generates different cascades of containment. The obserpant is subjected to constraints of both the microscopic and the macroscopic world, as he or she strives to decomplexify the world in order to enable smooth navigation. Thus, the fractal geodesics of spacetime constrain, as an ontological entity, the microscopic movements within us as well as those in our environment. At the same time, as a result of our co-evolution with a phase of the universe, our embodied brains have come up with the concept of fractal spacetime. As obserpants, we simultaneously partake in both the quantum and classical levels with which we co-evolved. It seems as if phenomenology and ontology are inextricably interwoven in our interfaces [50].

12.4 Outlook

Both local and global strong anticipation can be a blessing when they help us to coordinate with our environment and navigate the world smoothly by compensating distances and delays or coordinating internal and external long-term correlations. Epistemologically speaking, however, they are blind spots which constrain our perspectives. Revealing delays and distances which we have compensated unwittingly is a nontrivial undertaking. So is the coordination of embedded long-term correlation with embedded ones. In order to become aware of the transparent phenomenal and epistemological scaffolding we have acquired, we need to expose habits and natural laws which might result from our compensatory acts. Possibly, looking for compensated delays and distances can be practised by removing existing delays and distances or inserting delays and distances into our interaction with our environment. It may well turn out that habits and natural laws have evolved from our compensatory acts. Ridding ourselves of strong anticipation may manifest itself in direct perception, insight and nonlocal perspectives we would have to get used to. However, the price to pay for enlightenment, the removal of blind spots, or at least a revealing perspective, would be the end of naïve realism and smooth navigation.

References

[1] I. Newton: Four Letters … to Doctor Bentley, containing some arguments in proof of a deity (in reply to inquiries made by him before publishing the last two lectures of a confutation of atheism). *Nabu Public Domain Reprints Publisher.* Reproduction of the original letters. Letter III, pp. 25–26.

[2] C. W. Misner, K. S. Thorne and J. A. Wheeler, *Gravitation*, Freeman and Company, 1973, p. 5.
[3] J. A. Wheeler, Time today. In: J. J. Halliwell, J. Perez-Mercader and W. H. Zurek (eds.) *Physical Origins of Time Asymmetry*, Cambridge University Press, Cambridge 1994, p. 8.
[4] A. Einstein, B. Podolsky and N. Rosen, Can quantum-mechanical description of physical reality be considered complete? *Physical Review*, 47, 777–780, 1935.
[5] F. Levi, Nonlocal legacy. *Nature Physics*, 11, 384, May 2015.
[6] *ibid*, p. 384.
[7] O. E. Rössler, Relative state theory: Four new aspects. *Chaos, Solitons and Fractals*, 7(6), 845–852.
[8] H. Everett III, Relative state formulation of quantum mechanics. *Reviews of Modern Physics*, 29(3), 454–462, 1957.
[9] J. A. Wheeler, The anthropic universe (Science Show, abc.net.au). 18 February 2006.
[10] E.g.: R. Wagner, W. Kersten, H. Lemmel, S. Sponar and Y. Hasegawa, Quantum causality emerging in a delayed-choice quantum Cheshire Cat experiment with neutrons. *Nature. Scientific Reports*, 13, 3865, 2023.
[11] J. A. Wheeler, Time today. In: J. J. Halliwell, J. Pérez-Mercader and W. H. Zurek (eds.) *Physical Origins of Time Asymmetry*, Cambridge University Press, Cambridge, 1994, pp. 15–16.
[12] D. Bohm, *Wholeness and the Implicate Order*, Routledge, New York, 1980.
[13] J. S. Bell, Quantum mechanics for cosmologists. In: C. Isham, R. Penrose and D. Sciama (eds.) *Quantum Gravity*, Vol. 2, Clarendon Press, Oxford, 1981, pp. 611–637.
[14] O. E. Rössler, Relative state theory: Four new aspects. *Chaos, Solitons and Fractals*, 7(6), 846.
[15] *ibid*, p. 850.
[16] O. E. Rössler, *Endophysics*, World Scientific, Singapore, 1998; O. E. Rössler, Intra-observer chaos: Hidden root of quantum mechanics? In: M. S. El Naschie, O. E. Rössler and I. Prigogine (eds.) *Quantum Mechanics, Diffusion and Chaotic Fractals*, Pergamon Press, 1995, p. 105; O. E. Rössler, personal communication, 2007.
[17] O. E. Rössler, Descartes' Traum. Von der unendlichen Macht des Außenstehens. Audio-CD. *Supposé*, Cologne, 2002.
[18] A. Legget, Low-temperature physics, superconductivity and superfluidity. In: P. Davies (ed.) *The New Physics*, Cambridge University Press, 1989,

pp. 268–288; O. E. Rössler, Relative state theory: Four new aspects. *Chaos, Solitons and Fractals*, 7(6), p. 850.

[19] O. E. Rössler, *Das Flammenschwert oder Wie hermetisch ist die Schnittstelle des Mikrokonstruktivismus?* Benteli Verlag, Bern, 1994.

[20] M. Nadin, Blog archive anticipation — A spooky computation. In: *Contribution to the Conference on Computing Anticipatory Systems (CASYS 99)*, Liege, Belgium, August 8–11, 1999, p. 9. www.nadin.ws/archives/39. Accessed 23.5.23.

[21] *ibid.*

[22] "The entanglement swapping protocol entangles two remote photons without any interaction between them. Each of the two photons belongs initially to one of two independent entangled photon pairs (e.g., photons 1 and 4 of the entangled pairs 1–2 and 3–4). The two other photons (2 and 3) are projected by a measurement onto a Bell state. As a result, the first two photons (1 and 4) become entangled even though they may be distant from each other." (E. Megidish, A. Halevy, T. Shacham, T. Dvir, L. Dovrat and H. S. Eisenberg, Entanglement between photons that have never coexisted. *Physical Review Letters*, 110(21), 210403, 1, May 2013.

[23] *ibid*, p. 2.

[24] I. Marcikic, H. de Riedmatten, W. Tittel, H. Zbinden and N. Gisin, Long-distance teleportation of qubits at telecommunication wavelengths. *Nature*, 421, 509–513, 2003; R. Ursin, T. Jennewein, M. Aspelmeyer, R. Kaltenbaek, M. Lindenthal, P. Walther and A. Zeilinger, Quantum teleportation across the Danube. *Nature* 430, 849, 2004.

[25] S. M. Korotaev, V. O. Serdyuk and J. V. Gorohov, Forecast of geomagnetic and solar activity on nonlocal correlations. *Doklady Earth Sciences*, 415A(6), 975–978, 2007.

[26] S. M. Korotaev, N. Budnev, V. O. Serdyuk, E. Kiktenko, D. Orekhova and J. Gorohov, Macroscopic nonlocal correlations by new data of the Baikal experiment. *Journal of Physics: Conference Series* (Vigier Centenary, IOP Publishing), 2197, April 2022. Art. 012019.

[27] "Nonlocal correlations are manifested in the nanoscale entropy change of the liquid phase of the double electric layer (…). The detectors represent matched pairs of weakly polarized metrological silver-silver chloride electrodes HD-5.519.00 with a small (several centimeters) distance between contact windows." (S. M. Korotaev, N. Budnev, V. O. Serdyuk, E. Kiktenko and J. Gorohov, Macroscopic nonlocal correlations in reverse time by data of the Baikal experiment. *Journal of Physics: Conference Series 2020.*

XXI International Meeting of Physical Interpretations of Relativity Theory, IOP Publishing, Vol. 1557, p. 2, Art. 012026.

[28] S. M. Korotaev, N. Budnev, V. O. Serdyuk, E. Kiktenko, D. Orekhova and J. Gorohov, Macroscopic nonlocal correlations by new data of the Baikal experiment. *Journal of Physics: Conference Series* (Vigier Centenary, IOP Publishing), 2197, April 2022, Art. 012019.

[29] *ibid*, p. 10.

[30] *ibid*, p. 10.

[31] J. G. Cramer, Generalized absorber theory and Einstein-Podolsky-Rosen paradox. In: *Physical Review D*, 22, 362–376, 1980.

[32] N. A. Kozyrev and V. V. Nasonov, *Manifestation of Cosmic Factors on the Earth and Stars*. In: A. A. Efimov (ed.), VAGO Press, Moscow-Leningrad, 1980, pp. 76–84.

[33] N. A. Kozyrev, *Possibility of Experimental Study of Properties of Time*, Pulkovo, September 1967. https://www.astro.puc.cl/~rparra/tools/PAPERS/kozyrev1971.pdf.

[34] D. G. Stephen and J. A. Dixon, Multifractal cascade dynamics modulate scaling in synchronization behaviours. *Chaos, Solitons & Fractals* (Elsevier), 44(1–3), 160–168, 2011; V. Marmelat and D. Delignières, Strong anticipation: Complexity matching in inter-personal coordination. *Experimental Brain Research*, 222, 137–148, 2012.

[35] J. Krishnamurti and D. Bohm, *The Ending of Time*, Harper, San Francisco, CA, 1985.

[36] M. J. Hove, J. Stelzer, T. Nierhaus, S. D. Thiel, C. Gundlach, D. S. Margulies, K. R. A. Van Dijk, R. Turner, P. E. Keller and B. Merker, Brain network reconfiguration and perceptual decoupling during an absorptive state of consciousness. *Cerebral Cortex*, 26, 3116–3124, 2016.

[37] P. Weibel, The noise of the observer. In: T. Druckrey (ed.) *Ars Electronica: Facing the Future*, MIT Press, Cambridge.

[38] This section is based on my article on time and timelessness in Eastern and Western thought traditions. (S. Vrobel, The path to timelessness: Insight, assignment conditions and strong anticipation. *Progress in Biophysics and Molecular Biology* (Elsevier), 131, 162–170, 2017.)

[39] S. Vrobel, Sheer simultaneity. In: *IIAS Conference Proceedings*, Tecumseh, Canada, 2006, pp. 2–9.

[40] S. Erimaki, K. Kanatsouli, V. Tsirka, E. Karakonstanaki, V. Sakkalis, M. Vourkas and S. Micheloyannis, EEG responses to complex fractal stimuli. *Poster Session at the 2nd International Nonlinear Sciences Conference, SCPLS*, Heraklion, Greece, 2005.

[41] See, for instance: C. M. Hagerhall, T. Laike, R. P. Taylor, M. Küller, R. Küller and T. P. Martin, Investigations of human EEG response to viewing fractal patterns. In: *Perception*, 37(10), 1488–1494, 2008; R. Taylor, Reduction of physiological stress using fractal art and architecture. *Leonardo*, 39(3), 245–251, June 2006; see also Chapter 11.

[42] M. Gladwell, *Blink — The Power of Thinking Without Thinking*, Penguin, London 2006.

[43] R. Penrose, *The Emperor's New Mind*, Vintage, London, 1989, p. 575.

[44] S. Vrobel, *Fractal Time*, The Institute for Advanced Interdisciplinary Research, Houston, 1998.

[45] L. Nottale, Scale relativity, fractal space-time and morphogenesis of structures. In: H. H. Diebner, T. Druckrey and P. Weibel (eds.) *Sciences of the Interface*, Genista, Tübingen, 2001, pp. 38–51.

[46] S. Vrobel, *Fractal Time*, The Institute for Advanced Interdisciplinary Research, Houston, 1998.

[47] The PSC comes in many guises: Natural constraints on possible embeddings can be generated both by physical structures of the outside world and the obserpant's internal differentiation. Constraints manifest themselves in the form of embedding forces, such as gravity, and embedded ones, such as the limitations of our physical strength or our perceptual apparatus. It seems likely that also temporal natural constraints are selection effects.

[48] E. Husserl, *Vorlesungen zur Phänomenologie des inneren Zeitbewußtseins*, Niemeyer 1980 (First published in 1928).

[49] M. S. El Naschie, Young double-slit experiment, Heisenberg uncertainty principle and correlation in Cantorian space-time. In: M. S. El Naschie, O. E. Rössler & I. Prigogine (eds.) *Quantum Mechanics, Diffusion and Chaotic Fractals*, Elsevier Science, Pergamon, 1995, pp. 93–100; M. S. El Naschie, A review of E-infinity theory and the mass spectrum of high energy particle physics. *Chaos, Solitons and Fractals* (Elsevier Science), 19(1), 209–236, 2004.

[50] This section is based on an article on the philosophical nature of fractal spacetime. (S. Vrobel, Fractal time and fractal spacetime: Phenomenology or ontology? In: J.-H. He, *Fractal Spacetime and Noncommutative Geometry in Quantum and High Energy Physics*, Vol. 1, No. 1, Asian Academic Publisher Limited, Hongkong, China, 2011, pp. 55–58).

Index

Printed in the United States
by Baker & Taylor Publisher Services